Abortion and the Moral Significance of Merely Possible Persons

Philosophy and Medicine

VOLUME 107

For other titles published in this series, go to
www.springer.com/series/6414

Melinda A. Roberts

Abortion and the Moral Significance of Merely Possible Persons

Finding Middle Ground in Hard Cases

Springer

Melinda A. Roberts
Department of Philosophy and Religion
College of New Jersey
Ewing, NJ
USA
robertsm@tcnj.edu

ISBN 978-90-481-3791-6 e-ISBN 978-90-481-3792-3
DOI 10.1007/978-90-481-3792-3
Springer Dordrecht Heidelberg London New York

Library of Congress Control Number: 2010922448

Printed on acid-free paper

Springer is part of Springer Science+Business Media (www.springer.com)

Acknowledgements

For their comments on earlier drafts of parts of this book, and the useful discussions they've engaged in with me regarding many of the topics I address here, I am deeply grateful to Gustaf Arrhenius, John Broome, David Enoch, Fred Feldman, Axel Gosseries, Caspar Hare, Elizabeth Harman, David Heyd, Jeff McMahan, Alan McMichael, Ingmar Persson, Peter Singer, Larry Temkin, Peter Vallentyne and David Wasserman. Of special value were the comments I received in connection with the conference on Extensions of Justice, Hebrew University of Jerusalem (June 2009).

Contents

Chapter 1
Introduction

1.1 Goals

1.1.1 I have two main goals in this book. The first is to give an account of the *moral significance of merely possible persons* – persons who, relative to a particular circumstance, or possible future or *world*, *could* but in fact *never do* exist.[1] I call that account *Variabilism*. My second goal is to use Variabilism to begin to address the problem of abortion.

1.1.2 We ought to do the best we can *for people*. And we consider this obligation to extend to people who are, relative to a world, *existing* or *future*. But does it extend to *merely possible people as well*? And, if it does, then does it extend to making things better for them by way of *bringing them into existence*? If we say that surely it *doesn't*, does that then mean that our obligation to do the best we can for people does *not*, after all, extend to the merely possible – that the merely possible do *not* matter morally? But if the merely possible do *not* matter morally, then doesn't that mean that it would be permissible for us to bring them into miserable existences – and even *obligatory* to do just that – in the case where bringing the merely possible into miserable existences creates additional wellbeing for existing

[1]References to *merely possible persons* and, later on, to persons who do exist – *existing persons* – and persons who will exist – *future persons* – can succeed only if they are understood in relation to particular circumstances, or possible futures or *worlds*. Thus, a person may be *merely possible* (or *future* or *existing*) relative to one world but not relative to another. Accordingly, I relativize those terms to worlds (at least implicitly) throughout this work.

Some philosophers would rather not talk about the merely possible as though they were alive and well and sleeping on the couch in the next room (and would that they would leave). The concern is that such talk will predispose us to think that the merely possible matter morally. But talk about the merely possible is not *inherently* question-begging. At least, it does not beg any question of interest here. And making talk about the distinct ways in which the future might unfold and the distinct populations that do or will exist within those futures – within those *worlds* – out of bounds also makes it very cumbersome to say what we *do* need to say about the merely possible. It would be like trying to say *exactly* what it *means* to say *all men are mortal* without availing oneself of the quantifier. See, for example, Russell (1903), pp. 18–23.

M.A. Roberts, *Abortion and the Moral Significance of Merely Possible Persons: Finding Middle Ground in Hard Cases*, Philosophy and Medicine 107,
DOI 10.1007/978-90-481-3792-3_1, © Springer Science+Business Media B.V. 2010

and future people? Doesn't that mean that doing anything *less* than our best for existing and future people is *always wrong* – even when the reason we do less than our best for them is to avoid imposing miserable existences on the merely possible? But if we recognize *that* as a result that we cannot accept, then doesn't that mean that we are compelled, after all, to conclude that our obligation to do the best we can for people extends to the merely possible?

Breaking out of this tail-biting line of questioning requires some work. In this book, I try to do that work by developing a clear and plausible account of the moral significance of merely possible persons. That account – *Variabilism* – rejects the idea that the merely possible *always* matter morally. But it also rejects the idea that they *never* matter morally. Instead, it claims that the merely possible matter morally but also that they matter *variably* – and that that, as it happens, is *exactly* how all of the rest of us matter morally as well.

The cases that we examine in what follows suggest just such a variability. As we shall see, however, strictly speaking that variability extends not across the domain of *persons* but rather across the domain of *diminutions in wellbeing*, or *harms*, that persons, whether existing, future or merely possible, suffer – that is, across the domain of *losses* that they incur. Accordingly, we don't say that some *people* matter morally and others do not. Instead, where *loss* is understood as shorthand for any case in which agents could have created more wellbeing for a given person and instead create less, we say that some *losses* incurred by that person matter morally and others do not. Some losses incurred by a given person have, that is, *moral significance* for purposes of evaluating the acts that impose those losses and their alternatives and others do not.

The variability that the cases suggest is, moreover, not at all random. There is a pattern. The moral significance of any particular loss is a function of just *where* that loss is incurred in relation to the person who incurs it. More specifically: a loss incurred at a world where the person who incurs that loss does or will exist, according to Variabilism, has *full moral significance*, while a loss incurred by that very same person at a world where that person never exists at all has *no moral significance whatsoever*.

1.1.3 My second goal is to begin to address the problem of abortion. It is useful, I suggest, to explore abortion as a problem in population ethics.

For purposes here, I use the term *early abortion* to describe the abortion that leaves a person out of existence altogether. And I use the term *late abortion* to describe the abortion that ends the life of an already-existing person. An early abortion, in other words, imposes a loss on a person who never exists at all at the world where the early abortion is performed. It imposes a loss on a *merely possible person*. In contrast, a late abortion imposes a loss on a person who has already come into existence at the world where the late abortion is performed. It imposes a loss on an *existing* person. What makes the terms *early* and *late* abortion apt is a certain assumption – that *thinking* and *being a person* come together – in combination with certain empirical facts about just when during the process of development the human organism begins to think. What gives the distinction between the early abortion and the late abortion its moral relevance is a second assumption. The second assumption is that being a *person*, that is, a *thinking thing*, and being the kind of thing in respect of which we have moral obligations (*mattering morally*) themselves come together. What makes this second assumption plausible may simply be that it is in the case of thinking things that it *matters to them* (whether they know it or not) that we treat them in one way rather than another.

Certainly, these assumptions raise questions that warrant books of their own. For purposes here, however, I am interested in another set of questions. First, I want to know whether the loss incurred by the merely possible person when that person is left out of existence altogether *counts against* the choice of early abortion. Can, in other words, that loss make the *otherwise permissible early abortion wrong*? And, if we say it can't, I want to know, second, whether we can take that position without committing ourselves to the (highly problematic) further position that the merely possible do not matter morally *at all*.

Citing Variabilism, my argument will be that the early abortion, in any ordinary case, is perfectly permissible if it is what the pregnant woman herself wants, and that it can be determined to be permissible under a *very* summary moral analysis. So, similarly, can we determine that the choice not to conceive a child, or to conceive a child in vitro but not arrange for the newly created embryo to develop *in utero* into a person, is permissible in any ordinary case. For, according to Variabilism, the effect the early abortion has on the person it relegates to the class of the merely possible – the loss of wellbeing the merely possible person incurs when that person is never brought into existence to begin with – has *no moral significance whatsoever* for purposes of evaluating the early abortion or its alternatives. We can then simply observe that any otherwise plausible permissibility theory will imply, in any ordinary case, that the early abortion is permissible.

As we shall see, other accounts as well suggest very similar results for this very same class of cases. But those accounts take a radically *exclusive* approach in connection with the merely possible. They make the mistake of asserting that the merely possible person does not matter morally *at all*. And that mistake, in turn, leads the exclusive approach to generate a slew of results that are highly counterintuitive or conceptually problematic or both. Variabilism, in contrast, by acknowledging that the merely possible *do* matter morally – if variably – avoids those results. Yet Variabilism avoids those results without adopting a radically *inclusive* approach in respect of the merely possible. Variabilism thus retains the ability to explain just why the loss incurred by the merely possible person when we choose to leave that person out of existence altogether cannot count against that choice.

I will also be interested here in the problem of late abortion. Variabilism, I will argue, lends support to the view that, ordinarily, late abortion is wrong. The loss that is imposed by the late abortion, according to Variabilism, has *full moral significance*. It's a loss, after all, that is incurred by an *existing* person. And it's often a very great loss – the loss of the person's entire future life. Of course, there are morally significant losses on the other side as well – the pregnant woman and others may well incur losses if the pregnancy is continued. In such cases, the late abortion constitutes a *tradeoff scenario*. Those are always a little complicated, and the analysis cannot be quite as summary as it is in the case of early abortion. Still, in the case of late abortion, the tradeoff is often relatively simple. Under one choice, one person faces a very great morally significant loss. Under the other choice, other persons face losses that have full moral significance but are not, in any ordinary case, nearly so great. The upshot is that, in any ordinary case, the late abortion is going to be very hard to justify. Once the extent of the loss and its moral significance are both recognized, the otherwise plausible permissibility theory can be expected to declare the late abortion wrong in many cases.

As noted, these results are subject to the various assumptions about what counts as a person. The first assumption is that being a *person* and being a *thinking thing* are connected. I take that assumption to mean that, toward the beginning of the pregnancy, the human embryo or fetus, because it is not a thinking thing, is not a person. But late in the pregnancy, the human fetus, because it is a thinking thing, is a person. At that stage, moreover, it seems plausible that the fetus has acquired enough of a *capacity to survive* from one moment to another that the late abortion's *not* being performed would make things better for that person and not just remove someone from existence whose going out of existence was imminent in any case. The second assumption is that being a person and mattering morally – if variably – come together.[2]

These assumptions, together with Variabilism and our independent sense of how the morally significant loss *bears on* the question of a particular act's permissibility, make the question of *when* the developing human (or non-human) organism begins to think in a way that is critical to that thing's mattering morally very important for purposes of evaluating any given abortion choice. According to the rough timeline I will later describe, the *thinking* that signals the coming into existence of a person, along with the *capacity to survive* that means that the abortion's *not* being performed would make things better for that person, does not take place until at least half-way through the pregnancy.[3] Moreover, the period of time between the moment at which we have *thinking plus the capacity to survive*, and the moment of *birth*, may well itself

---

[2] We can leave open whether that obligation itself is rooted in an obligation to make things better for certain things *for their own sake* or an obligation to make things better for certain things *for the sake of the universe*. See part 1.3.3 below.

[3] Of course, much depends on what we mean here by *thinking*. But it seems clear that mere electrical activity is not enough. Nor are mechanical responses to pain stimuli. See Peter Singer (1994), pp. 104–105. At the same time, for purposes here we want a concept of *thinking* that is broad enough to extend to conscious beings whose thoughts are quite unsophisticated. They need not have a life plan or a clear concept of themselves as continuing beings. Moreover, we should recognize that thinking often takes place beyond our own conscious awareness. I thus take it for granted that the cognitive activity and degree of continuity we typically see in both the human neonate and in many non-human animals is enough to show that the destruction of those individuals is a matter of ending the life of an already-existing *person* and not the choice *not* to bring a new person into existence to begin with. More generally, the guide for what counts as *thinking* for purposes here is the idea that the kinds of things that matter morally – that are *persons* – are the kinds of things such that it *matters to them* (whether they think so or not) whether we treat them in one way rather than another. If there's enough cognitive activity that that latter is the case, then that activity should be considered thinking, for purposes here. The connections among being a person, mattering morally and thinking are discussed in more detail in parts 1.6.5, 5.2 and 5.3 below.

be substantial – perhaps weeks in duration if the pregnancy is full-term. The timeline I provide, however, is rough – and it's really only a good guess. For my aim in this book is to get the moral significance of the *merely possible person* as it relates to abortion exactly right. It is not getting the timeline itself exactly right.

1.1.4 The matters I have dealt with by assumption here are examined in many old debates on abortion in considerable depth. I think the assumptions I make here are highly plausible. However, even if they are accepted as fully *correct*, on their own they will not even begin to resolve the newer – or at least off-grid – debate having to do with abortion that I want to address here.

That newer debate is triggered by what is really just an argument *against* early abortion – and non-conception and conception followed by non-fruition. The argument is this. Suppose that we become convinced by the cases that it is false that merely possible persons don't matter morally and neither do any of their losses. Suppose that we become convinced that it cannot really be wrong for us to impose some slight loss on existing or future people in a case where the only way we can make things better for them is to bring one or more merely possible persons into existence and treat each such person badly. We will then seem compelled to say as well that the losses that merely possible persons incur are, after all, morally significant. We seem compelled to say that those losses *bear on* the permissibility of the acts that impose those losses and the alternatives to those acts that avoid those losses. And we seem in turn compelled to say as well – by simple universal instantiation – that the loss the merely possible person incurs when that person is never brought into existence to begin with is morally significant and must be *counted against* the early abortion.

Now, it's true that this position does not absolutely preclude the result, in a given case, that the early abortion is permissible. But the analysis will be arduous and, it seems, the result often will be that the early abortion is wrong. After all, the loss that *counts against* the early abortion will be so much graver than anything on the other side that might count *in its favor*.

Variabilism challenges this argument. According to Variabilism, some of the losses incurred by the merely possible have no moral significance at all while still others have full moral significance. We can say that the loss the merely possible person would incur, were we to bring that person into existence and treat that person badly, has full moral significance and counts against that choice. For that loss would be incurred at a world where that same person does or will exist. And we can say at the same time that the loss that same person incurs as a result of the early abortion has no moral significance whatsoever. For that loss is incurred at a world where that person never exists at all.

A variable treatment thus justifies the summary analysis in favor of early abortion – and non-conception, and conception followed by non-fruition. But it does so without putting any pressure on us at all to declare ourselves *obligated* to bring the merely possible into existence and then proceed to treat them in outlandishly bad ways.

1.1.5 The question of the moral status of merely possible persons and the question of the permissibility of early abortion are thus closely connected. I am ordinarily not a big fan of the middle ground. But Variabilism undeniably occupies a middle ground between two very extreme positions, one that *radically includes* and another that *radically excludes* the merely possible among those who matter morally. According to Variabilism, each person – existing, future or merely possible – incurs *some* losses that have moral significance and *others* that don't, with the moral significance of any loss being a function of *where* that loss is incurred in relation to the person who incurs it. More specifically, a loss incurred by a person at a world is morally significant for purposes of evaluating the act that imposes that loss and each of that act's alternatives *if* that person does or will exist at that world. But the loss incurred by any person – including you and me – at any world where that person never exists at all is completely devoid of any moral significance whatsoever.

This middle ground on the moral significance of merely possible persons opens the door to a middle ground on abortion. Those who think abortion is morally permissible throughout the pregnancy are surely most concerned to have that fact recognized in connection with *early abortion*. And those who disdain abortion at any stage of pregnancy for any reason are surely most concerned to see abortion recognized as impermissible late in pregnancy. My proposal reflects both those concerns. It, in effect, asks each of the two camps to settle for the ground that is most important to it and concede the ground that is most important to the opposition.

1.2 Organization of Book

1.2.1 Do *merely possible* persons matter morally? In evaluating any given act for its permissibility, is a *part* of what is important how the merely possible are affected by that act and its alternatives? Or, alternatively, are persons who *do* or *will* exist the only persons who matter morally? Is it just their needs and interests we must take into account in calculating what we ought to do?

Why these are interesting and important questions, in their own right and in connection with abortion, is outlined in this Chapter 1. My suggestion will be that we can think most productively about these questions if we think not in terms of *who* matters morally but rather in terms of *what losses* matter morally. We can, accordingly, rephrase the critical questions. Do the *losses* the merely possible incur when we leave them out of an existence worth having *count against* the acts that impose those losses? And do those same losses, in a roundabout way, *count in favor of* the alternative acts that *avoid* them?

1.2.2. In Chapter 2, I examine a handful of competing rules on just when the losses incurred by any merely possible person have moral significance and when they do not. Now, those rules do not, on their own, generate any *permissibility* results at all. Rather, each rule simply generates a collection of *moral data* consisting of certain facts that, according to the rule, *bear on* permissibility. Each rule, in other words, does no more than offer a competing picture of which losses are to be taken into account in determining what we ought to do and which losses are *not* to be taken into account in determining what we ought to do.

But, for purposes of testing, it is permissibility that we often have some clear ideas about. And, at least eventually, it is permissibility results that we will need for purposes of giving an account of abortion. Accordingly, in order to test the rules and apply the rule we like to the case of abortion, we will need permissibility results. We will need to see how a given rule on when losses have moral significance (a *loss rule*) combines with principles (*permissibility principles*) that instruct, on the basis of data generated by that rule, which acts are permissible and which are not.

Happily, we can articulate a handful of reasonably plausible permissibility principles that we can accept *regardless* of our views on just which loss rule is correct. These principles together make up what I will call the *otherwise plausible permissibility theory*: the theory that, but for whatever loss rule we happen at the moment to have combined that theory with for purposes of testing, is otherwise plausible.

The first order of business in Chapter 2 is thus to describe that theory, which is also summed up in Appendix A. I shall also need, starting out, to address certain other preliminary matters. For example, before we can say whether a given loss is morally significant or not, we need to say something about loss. For purposes here, I accept a *maximizing* account of loss – an account that finds *loss* whenever agents (by act or omission) create less wellbeing for a person when they could have created more wellbeing for that same person.[4] A bit more needs to be said,

[4] The maximizing account of loss I adopt for purposes here is described in parts 2.2.1–2.2.3 below. Briefly, my account provides that a person – existing, future or merely possible – incurs a *loss* whenever agents (by act or omission) create less wellbeing for that person when they *could have created more* wellbeing for *that same person*. To say that a person incurs a loss is not to say that that person has been wronged, or that a wrong has been done.

This maximizing, comparative account of loss rejects both a *temporal* approach to loss (a person incurs a loss *only if* that person is worse off at a later than at an earlier time) and a *counterfactual* approach (a person incurs a loss *only if* that person would have been better off had the act under scrutiny not been performed). It also rejects a *non-comparative* approach to loss – an approach that defines loss by reference either to a list of specific bad effects (e.g. substantial pain) or to a threshold (e.g. a minimally decent wellbeing level). Finally, it accepts the concept of netting benefits generated by an act for a person against burdens. On my account, we cannot conclude, from the fact that burdens have been generated by the performance of a particular act, that that act imposes a *loss* on a person at a world. A vaccination can impose pain, but that the vaccination imposes some pain does not mean that the vaccination imposes a loss; pain can be imposed even in a case where wellbeing has been maximized.

I also accept, for purposes here, that one way of causing a person to incur a loss – one way, that is, of creating less wellbeing for a person rather than more – is to leave that person out of an existence worth having. Some theorists take the position that a person *cannot* incur a loss at a world where that person *never exists at all*. It seems to me, however, that the concerns that lead to that position can be overcome, and that comparisons between a person's wellbeing level at a world where that person never exists at all against that same person's wellbeing level at a world where that person does or will exist are fully cogent. For a brief discussion of this issue and references to work in this area, see part 2.2.2 and note 45 below.

Some theorists might think that the *nonidentity problem* means that there will be many cases (A) where we *want* to say – and where naively think we *can* say and *need* to be able to say, if we want to make loss central to an account of wrongdoing – that a loss *bears on* the permissibility of a given act but (B) that in fact involve no harm, or loss, at all. We can see how agents could have brought a *nonidentical* person into a better existence, but we cannot – or so the nonidentity problem asserts – see how agents could have brought that *very same person* into a better existence.

however, especially in light of the fact that, to avoid begging any important questions, we shall need an account of loss that recognizes that leaving someone out of an existence worth having is to impose a *loss* on that person. Only then can we open-mindedly ask the next question: is that loss morally significant?

With the preliminaries out of the way, we then examine, also in Chapter 2, the competing rules on when losses incurred by the merely possible have moral significance and when they do not.

The loss rule that I will call *Inclusion* proposes a *radically inclusive position* regarding which losses are morally significant and which are not. *Classic Utilitarianism* (*Totalism*) is a paradigm example of a view that adopts Inclusion. On that view, agents are obligated to make the choice that maximizes wellbeing on a *total*, or *aggregate*, basis. Whether they do that by creating additional wellbeing for existing or future people, or by creating additional well-off people, is immaterial. The upshot is that the loss a person incurs when that person is left out of existence altogether has full moral significance.

According to Inclusion, the merely possible matter morally just as you and I do, and all of the losses they incur, just like all of the losses you and I and each other existing and future person incur, have moral significance for purposes of evaluating the acts that impose those losses and the alternative acts that avoid those losses.

I argue that the cases show that Inclusion, in combination with the otherwise plausible permissibility theory, is completely untenable. Inclusion fails to take into account a critical, intuitive moral distinction that can only be described as axiomatic. Inclusion must, accordingly, be rejected.

Still another loss rule, what I will call *Exclusion*, is, when combined with the otherwise plausible permissibility theory, roughly equivalent to the view that is sometimes called *Moral Actualism*. Exclusion proposes a *radically exclusive position* regarding loss. According to Exclusion, *merely possible* persons and *existing*

In this book, I discuss the nonidentity problem only briefly, in part 2.2.5 below and in Appendix B. For purposes here, I mainly set it aside, in part because I have discussed it in detail elsewhere (e.g. Roberts 2009c, 2007) and in part because I believe that – *properly understood* – it undermines *nothing* we want or need to say here. My basic point is that the type of nonidentity problem considered among the most telling relies on a fallacy. Cases where I argue that that is so include (among others) Parfit's depletion and risky policy examples, Kavka's slave child and pleasure pill examples and cases involving historic injustices. Once we recognize the fallacy, I argue, we can and must recognize that losses have been incurred and harms imposed. Moreover, in the rare case where we are dealing with a type of nonidentity problem that validly establishes that the act under scrutiny harms and is bad for *no one* – where, for example, a genetically impaired child is brought into an existence that is unambiguously worth having and the choice to bring that child into existence is *maximizing* for that child *and* for each other existing or future person – we are also dealing with a type of nonidentity problem that never clearly establishes that a wrong has been done.

Thus, it seems to me that we should not be too concerned with the nonidentity problem in constructing our moral theory. It is *not* that I think that the choice of depletion or the risky policy, or the choice to enter into the slave child contract or to take the pleasure pill, is permissible. It's rather that we can discern harm, that is, *loss* (and in a comparative, intuitive sense of those terms) in each of those cases. I thus argue that we can resolve the nonidentity problem *without* making the merely possible matter morally *more* than they ought and *without* disconnecting wrongdoing from acting in ways that are bad for some person or another. The key is to take the various types of nonidentity problem one by one and not try to resolve them all in one blow.

or *future* persons are in very different moral boats. *They* don't matter morally, but *we* (existing and future persons) do. Accordingly, while the losses *we* incur have full moral significance, the losses *they* incur have none at all.

Now, as we shall see, Exclusion itself comes in two forms, what I will call its *Alpha* and *Beta* forms, with the Beta form representing an attempt to address some of the issues created by the Alpha form. (Moral Actualism, similarly, is said to come in two forms, *strong* and *weak*.)

I will argue, however, that Exclusion in both its forms is just as untenable as Inclusion is. Exclusion Alpha, in combination with the otherwise plausible permissibility theory, suggests that performance is a determinant of permissibility – that an act can be *permissible*, or even *obligatory*, if *unperformed* but *wrong if performed*.[5] And Exclusion Beta suggests that an act can be *permissible* when its alternative is *obligatory*.[6] Exclusion Alpha seems conceptually unstable while Exclusion Beta on the face of things seems to be flatly inconsistent.

The very gravity of the difficulties that we see in Exclusion may well make us wonder whether we were right to reject Inclusion to begin with.

But we were, for there is a *middle ground* between Inclusion and Exclusion – and thus no need to think that rejecting Inclusion means we are bound to accept Exclusion. According to that middle ground – *Variabilism* – we all – existing, future *and* merely possible – *matter morally*, but we all *matter variably*.

According to Variabilism, each person, whether existing, future or merely possible, incurs *some* losses that have moral significance and *others* that don't, with the moral significance of any loss being just a matter of *where* that loss is incurred in relation to the person who incurs it. That is:

> *Variabilism*:
>
> Any loss incurred in any circumstance, or possible future or *world*, by *any* person has moral significance for purposes of evaluating the act that imposes that loss or any alternative to that act *if and only if* the person who incurs that loss does or will exist *at the world at which that loss is incurred.*

Variabilism has some important features. The conceptual difficulties we see in Exclusion are rooted in its attempt to say that some people do not matter morally, and neither do any of their losses. Variabilism avoids those conceptual difficulties by taking the view that all persons, existing, future and merely possible, matter morally, and in exactly the same ways.

Inclusion has that last feature in common with Variabilism. But Inclusion then blinds itself to the distinction between (A) the loss the *merely possible* child incurs when we choose not to bring that child into existence to begin with and (B) the loss the real, live, flesh and blood, *existing* child incurs when we create less wellbeing for that child when we could have created more. Variabilism does not do that.

[5] Exclusion Alpha, in other words, violates Wlodek Rabinowicz's Principle of Normative Invariance. See note 61 below.

[6] This result violates a standard deontic axiom that asserts that if one act is obligatory, its alternatives cannot be permissible. See part 2.2.4 below.

According to Variabilism, while we all matter morally, we all matter variably. Thus, the merely possible child matters morally, but the loss that child incurs when he or she is left out of existence altogether has *no bearing* on the permissibility of what we have done and, accordingly, can't make the otherwise permissible act wrong. It's the loss in relation to where it is incurred, not the child, that makes that so. It's the loss being incurred at a world where the child never exists that makes it devoid of moral significance. The existing child (of course!) matters morally as well. And the loss that child incurs, when we create less wellbeing rather than more for that child at a world where that child exists, has full moral significance and can easily make the otherwise permissible act that imposes that loss wrong. But again it's the loss and where it is incurred, not the child, that makes that so.

Now, Variabilism does not exhaust the middle ground between Inclusion and Exclusion. A position that John Broome calls the *Neutrality Intuition* and a related position that Peter Singer calls the *Prior Existence View* both can be construed as occupying middle ground as well. The test for determining whether one choice is just as good as another that is proposed by the Neutrality Intuition sets aside the interests of any person who does or will exist under the one choice but not the other – the interests, that is, of any additional person whose coming into existence is at stake – provided that person's existence falls within a "neutral range." Similarly, the Prior Existence View asserts that the determination of the permissibility of any particular choice is to be made by reference *only* to how well off the persons are who do or will exist *independently* of that choice – thus again setting aside the interests of the additional person.

But Broome and Singer argue – correctly, I think – that the Neutrality Intuition and the Prior Existence View are unacceptable. What I want to point out for purposes here is just that Variabilism is a quite different view. We can, that is, easily reject the Neutrality Intuition and the Prior Existence View, as Broome and Singer argue that we must. But there will be nothing in those arguments that suggests that Variabilism is itself defective. Just the reverse: Variabilism helps us articulate how those views go awry.

The notion that the interests of the additional person are to be set aside for purposes of determining whether that person ought to be brought into existence or not is – like Exclusion – sometimes associated with what is called the "person-affecting," or "person-based," intuition. According to the person-based intuition, an act that is wrong must make things worse for, or harm, or impose a loss on, some existing or future person. The association, however, between the person-based intuition and Exclusion is an unfortunate one. Thus, the Neutrality Intuition regards the addition of a person to a world as very often "*morally neutral*."[7] So does the Prior Existence View. It is, as we shall see, that very fact that in the end shall mean that both views must be rejected. In contrast, Variabilism regards the addition of a person to a world

---

[7] Broome (2004), p. 143 (emphasis added). I here compress (in Broome's terms) the *deontic* reading and the *teleological* reading of the person-based intuition. One pertains to the evaluation of *acts*; the other to the two-place betterness relation that we may think obtains between *outcomes*, or *worlds*, or *acts* performed at worlds. See Broome (2004), pp. 140–149. The two constructions of the person-based intuition are prized apart in part 2.10 below and Appendix C.

as very often *morally hazardous*. According to Variabilism, such an addition cannot, on its own, make things better but it can make them worse: it can convert an act that would be deemed permissible under an otherwise plausible permissibility theory into an act that is wrong. Variabilism reaches a different result in virtue of the fact that it considers any loss the additional person incurs to have full moral significance provided that the person does or will exist at the world at which the loss is incurred. According to Variabilism, then, any loss that the person incurs at a given world at which that person has been "added" – the loss that person incurs, not relative to the world where that person never exists at all, but rather relative to some third world at which that person enjoys a *better* existence – would count against the choice to bring that person into existence to begin with. Variabilism, accordingly, is able to avoid the objections that force us to reject both the Neutrality Intuition and the Prior Existence View. And, as we shall see, Variabilism along the way generates an interpretation of the person-based intuition that retains the full force of that intuition but also makes the intuition far more plausible than the interpretation that is suggested by either the Neutrality Intuition or the Prior Existence View.

The view that Singer calls the Prior Existence View expresses an "intuitive judgment" that Singer suggests that many people probably have – and a judgment that Totalism, the main alternative view that Singer is at that point considering, must reject. According to that judgment, the ordinary couple, contemplating whether they ought to have children, need not take into account in making their decision the "likely future pleasure of their children." Their choice whether to have children or not should instead be made on the basis of the effects that choice will have on people who do or will exist *independently* of that particular choice.[8]

Singer argues, however, that still other cases – cases in which we anticipate not that the child will be happy but rather that the child will be miserable if he or she exists at all – show that the Prior Existence View fails. Singer's argument is persuasive. We must reject the Prior Existence View. However, for purposes here, what is important is that rejecting the Prior Existence View does not mean that we are bound, after all, to say that the "likely future pleasure" of the child must be taken into account in determining whether the couple is obligated to bring that child into existence.

Variabilism makes that point. It offers an analysis that explains just why the child's prospective misery must be counted against the choice to bring the one child into existence, but also why the child's prospective happiness need not be counted in favor of the choice to bring the distinct child into existence. The upshot is that the fact that we must reject the Prior Existence View does not mean that we must reject Variabilism as well – or that we are in the end compelled to agree that the loss incurred by the merely possible person when we leave that person out of existence altogether can make the otherwise permissible act wrong.

Singer's discussion of the Prior Existence View takes him into the very deep waters of what is called the *Asymmetry*. The two claims that constitute the Asymmetry are just these. The fact that a possible person's life would be well worth living does not count in favor of bringing that person into existence and cannot, on its own, mean that it is wrong not to do that. *But* the fact that a person's life would

[8] Singer (1999), pp. 103–104.

be completely miserable – *less* than worth living – counts against bringing that person into existence – and may well mean that it would be wrong to bring that person into existence.[9] Totalism, implausibly, denies the first of these two claims. That is what makes the Prior Existence View attractive to begin with. But, at least as implausibly, the Prior Existence View denies the second of these two claims. It is for that reason that we must reject the Prior Existence View.

Having rejected the Prior Existence View, we may feel that we are forced as well to reject the Asymmetry itself. We may feel we have no choice but to consider the additional person's prospective happiness – in my terms, the *loss* the merely possible person will incur if that person is left out of existence altogether – as *morally significant* for purposes of evaluating the choice whether to bring that person into existence or not.

Variabilism, however, shows that that is not the case. According to Variabilism, what is important is *where* the losses are incurred in relation to the person who incurs them. Since, for the former claim, the loss is incurred at a world where the person who incurs that loss never exists at all, that loss has no moral significance whatsoever. So it does not count against the couple's choice to impose that loss – their choice, that is, not to have children. In contrast, for the latter claim, the loss we are worried about – the loss the person incurs having been brought into an existence less than worth having – is incurred at a world where the person does, or will, exist. So that loss, according to Variabilism, has full moral significance: it counts against the choice to bring that person into existence.

Thus, Variabilism, by taking, in a perfectly principled way, one result from Totalism and one result from the Prior Existence View, comes up with a treatment of the Asymmetry that itself seems highly plausible.

1.2.3 In Chapters 3 and 4, I explore a handful of theories that aim to throw – or, even if they do not *aim*, still, if correct, and when put together with other principles and deontic axioms we may be hesitant to reject, *succeed* in throwing – the permissibility of early abortion into serious doubt. Those theories weigh in, to one degree or another, *against* early abortion. And each does so on the basis of its own separate and, I argue, mistaken conception of just how merely possible persons matter morally.

In Chapter 3, I examine Jeff McMahan's Pareto-inspired paradox of abortion and prenatal injury (what I will call the *Abortion Paradox*).[10] McMahan argues that, for a class of scenarios that on the face of things seem morally indistinguishable, some abortions are wrong and others are perfectly permissible. What is of particular interest here is the half of the paradox that argues *against* abortion.

McMahan himself proposes to preserve both halves of the paradox. And he proposes to do so by taking note of a distinction in the *order of presentation of alternatives to the agent*.

But we may not see *why* the *order of presentation* – in contrast with, for example, the unordered contents of the relevant *feasible set* – can make such a difference to permissibility. If the logic that McMahan appeals to functions as intended in one

---

[9] McMahan (2009), p. 49.

[10] See McMahan (2006) and Chapter 3 below.

case, it may seem that, whatever the order of presentation, that same logic should function in the same way in the second case as well. It may seem that we end up with the result that the early abortion is itself *wrong across the board.*

I argue that a more productive approach is to challenge the Abortion Paradox itself. In particular, I question McMahan's application of what may seem to be a perfectly plausible Pareto principle. I argue that this principle – which makes an appearance not just in the half of McMahan's paradox that argues that abortion is wrong but also in the *Mere Addition Paradox* – is in fact highly questionable.

Variabilism is useful in developing this critique. Where the abortion under discussion is an early abortion, Variabilism challenges the inference from (1) the wrongness of the choice to *injure* the fetus in a case where the fetus survives and the person it develops into is then itself injured as well, to (2) the wrongness of the choice not to bring that same person into existence to begin with. The inference may seem plausible, since from the perspective of the fetus and the potential person the second loss is deeper than the first. If imposing the first is wrong, surely imposing the second is wrong as well. Appealing to Variabilism, however, we can point out that the depth of a loss is one thing, and its moral significance another. And we can argue that the loss that is incurred by the person as a result of the fetal injury has *full moral significance*; it *counts against* the choice of fetal injury. But we can also argue that the loss incurred by the person who is never brought into existence to begin with is *devoid of moral significance* – and that loss, accordingly, cannot be counted against what has been done.[11]

1.2.4 Chapter 4 examines three additional positions – from R.M. Hare, Don Marquis and Elizabeth Harman – that, whether by design or not, place the permissibility of the early abortion in doubt. Each of the three positions seems to assign at least some moral significance to the loss incurred by a merely possible person when that person is left out of existence altogether. Those positions are, in turn, at odds with the account of the moral significance of loss that Variabilism itself suggests.

Hare – like anyone who thinks we ought to *maximize wellbeing in the aggregate*; like any Totalist – can be understood to say that the merely possible matter morally in a certain robust way. Hare's view can be put in terms of *loss*. According to him, all losses incurred by all merely possible persons, including the losses they incur when we fail to bring them into existence to begin with, have full moral significance for purposes of evaluating the acts that impose those losses and their alternatives. These moral data in hand, we can then expect the otherwise plausible permissibility theory to

[11] McMahan sees the later person as *numerically identical* to the earlier fetus, the fetus that is the subject of the hypothetical fetal injury. I adopt a distinct metaphysics for purposes here. The fetus, a human organism, may well be identical to the *human organism* that develops in utero, is born and then dies. On the way I am looking at things here, however, it's not identical to the *person*, a being whose coming into existence is (I suggest) signaled by its *first thought*. This way of looking at things, I believe, lends clarity to the discussion and has an independent plausibility. But whether I am correct or not on that point, it would be a mistake to think that the *metaphysics* is ultimately what is at issue here. A different metaphysics is not going to erase the substantive moral issues that are (I will argue) raised by McMahan's Paretian analysis. At most, it would force us to *rearticulate*, not to *concede*, the underlying moral issues. See notes 24 and 143 below.

conclude in any ordinary case that early abortion – and non-conception, and conception followed by non-fruition – is wrong. That is so, even taking into account that, according to Totalism, the loss incurred by the person who never exists at all as a result of the early abortion must be balanced against the losses incurred by the pregnant woman or others if the pregnancy is permitted to proceed – even when, that is, the calculation of aggregate wellbeing is done, as it should be under Totalism, *on a net basis*.[12]

Marquis whittles down the class of those who matter morally by embracing what we can call an *existence condition*. According to Marquis, we are to worry about the loss that is imposed on the person who never exists at all *only if* the human embryo, or early human fetus, that may ultimately develop into that person *in fact exists* and has a future of a certain kind – a "future like ours" or a "future of value."[13] Thus, where Hare seems committed to the view that non-conception is just as morally problematic as early abortion, Marquis's approach targets just abortion (early and late). On the other hand, since Marquis is no utilitarian – for him, it may well be wrong to deprive the merely possible person of its future of value by way of early abortion even if that early abortion happens to maximize aggregate wellbeing – his stand against early abortion is in this way more sweeping than Hare's.

Despite their differences, however, it is clear that Hare and Marquis have in common extravagantly generous attitudes in respect of the merely possible. For both, the losses incurred by persons who, by dint of the early abortion, *will never exist at all* have moral significance for purposes of evaluating the early abortion. But Inclusion, I will argue, whether restricted in the way that Marquis suggests or not is a mistake. It is a mistake to say that the losses imposed on merely possible persons as a result of our choice not to bring them into existence to begin with have moral significance. It's a mistake to assign to the merely possible any sort of *necessary* moral status – a mistake to say that *all* of their losses have moral significance. And it's a mistake that Variabilism tries to correct, by understanding that some losses incurred by the merely possible have full moral significance and others have none at all.

Elizabeth Harman aims to reconcile two positions. The first is that early abortion is permissible even when continuing the pregnancy would impose no burden on the pregnant woman herself or anyone else. The second is that some early fetuses but not others have a certain moral status, a status that gives agents a moral reason to bring it about that those fetuses but not others develop into full-fledged persons and that makes those fetuses but not others appropriate objects of their progenitors' love. The difference, for Harman, is between the fetus that – because the agents choose against early abortion – will develop into a person and the fetus that – because the agents choose in favor of early abortion – will not. In one case, in other words, we are dealing with a person who does not yet but will exist, and in the other case we aren't.

Thus Harman, like Marquis, adopts a certain *existence condition*. But unlike Marquis's, Harman's is a forward-looking existence condition. It is a condition that

[12] See Hare (1973) and part 4.3 below.

[13] See Marquis (2008) and part 4.4 below.

is satisfied not by the mere existence of the *human embryo* or *early fetus* but rather only by the *eventual* coming into existence of the *person*. The *future person* has moral status, and hence *so does the existing fetus*. The *merely possible* person has no moral status, and hence *neither does the existing fetus*.[14]

The puzzle that is raised by Harman's approach is whether we must now say that the early abortion in fact performed is *permissible* but also that, at an alternate world where the agent chooses to continue the pregnancy, that choice becomes *obligatory*. This combination of results creates a serious conceptual difficulty. After all, if the early abortion is itself *permissible*, it seems that its alternative – continuing the pregnancy – cannot at the same time be *obligatory*.

Harman's view, by its own terms, seems to avoid any such result. On her view, the person who never exists at all in one world, where the early abortion is in fact performed, nonetheless has a certain *moral status* in an *alternate* world where the pregnancy is continued and the person eventually exists. Accordingly, agents at that alternate world have a *moral reason* to bring that person into an existence worth having. But from that fact we cannot, according to Harman, conclude that agents are there *obligated* to bring that person into existence. That is so, even in the case where agents have no moral reason not to bring that person into existence and no moral reason to do anything other than continue the pregnancy.

In short, by insisting on a certain disconnect between moral status and moral reason on the one hand, and moral obligation on the other, Harman avoids the conceptual problem.

It seems, however, that the proposed disconnect raises questions of its own. If we – unlike Harman – have a hard time prizing apart moral reason and moral obligation – if we accept that the fact that we have a moral reason to do something, and no moral reason not to do that thing, and no moral reason to do anything else, then we are obligated to do that thing – we are going to find in Harman's analysis either an inconsistency or, at best, a subtle argument *against* early abortion.

The better view, I suggest, is to begin with the view that the agent does not have a moral reason to continue the pregnancy – even at the world where that is exactly what the agent does, and where we agree that any losses incurred at that world by the person who will exist as a result of that choice will have full moral significance. Variabilism gives us a foundation for that position. According to Variabilism, the losses that have moral significance include just the losses that that person incurs at that world or any other where that person does or will exist. That means that the loss that person incurs at the world where the early abortion is in fact performed remains *devoid of moral significance*, even when the act we happen to want to evaluate – continuing the pregnancy – is performed at a world where that person does or will exist. We are thus left, according to Variabilism, with *no moral reason to explain away* – no moral reason that we must then demonstrate *not* to be connected in a natural way to our assessments of moral obligation.

The upshot is that each of these four theories, whether by the intention of their authors or otherwise, creates serious questions regarding the permissibility of the

[14] See Harman (2000) and part 4.5 below.

early abortion. Moreover, in each case, those questions arise out of the positions the particular philosopher seems to want to maintain regarding the moral significance of the merely possible person.

My own view is that we will obtain a more plausible account of early abortion when we have in hand a more plausible account of the moral significance of merely possible persons. And I think that Variabilism is just such an account. Variabilism proposes a *middle ground* between Inclusion and Exclusion. And that *middle ground on the merely possible* opens the door to a *middle ground on abortion*. Variabilism shows that we can *both* accept that the merely possible person has a moral status that is identical to our own – that is, a moral status *of sorts* – *and* take the position that the loss incurred by the merely possible person when we leave that person out of existence altogether has *no moral significance whatsoever* for purposes of evaluating the act that imposes that loss or any of that act's alternatives. The trick is to recognize the *variable moral status of each merely possible person – and each of the rest of us as well.*

1.2.5 The focus of Chapter 5 is abortion. I argue, with reference to Variabilism, that the choice of early abortion is ordinarily permissible and the choice of late abortion is ordinarily wrong.

How these results regarding moral permissibility are to be translated into the law – what they mean, if anything, for the law – is an important question. Both the equal protection clause of the Fourteenth Amendment of the U.S. Constitution and the due process clause are – as we shall see – critical in connection with this discussion.

Someone recently pointed out to me that what I call a *middle ground* on abortion is in fact going to offend practically everyone.[15] But I don't think it should. Those who disdain abortion at any stage of pregnancy for any reason are surely most concerned to see abortion recognized as impermissible late in pregnancy. And those who think abortion is permissible throughout the pregnancy are surely most concerned to have that fact recognized in connection with early abortion. My proposal reflects both those concerns. It asks each of the two camps to *take the ground that is most important to it – to settle for that – and concede the ground that is most important to the opposition.*

The discussion of early abortion may well remain fraught. But that may be less moral law and more a matter of evolution: a principle of genetic survival that leads many men – and no doubt many *grandmothers* as well, and let's not even talk about *grandfathers* – to find in moral law edicts commanding women to have babies. The discussion of late abortion may remain fraught as well. But there, too, we can identify a source of angst beyond moral law. We can recognize that there is something very wrong with a community that requires women to do so much for their soon-to-be-born offspring and allows everyone else to do so little for anyone at all. And we may well wonder whether one person's choice can be deemed wrong when everyone else is doing things that are at least as bad and

[15]I owe this point to Mark K. Greene.

getting away with them. In the end, however, it seems that these ancillary facts are not going to change the conclusion that we ultimately reach here – that the choice of late abortion is ordinarily wrong. That there's something wrong with the *community* does not imply anything at all about what is permissible for the pregnant women or any other *individual* agent to do.[16]

1.3 Inclusion, Exclusion and a Dilemma

1.3.1 Does the loss a merely possible person incurs when we leave that person out of existence altogether *count against* the choice that imposes that loss? Can that loss make the otherwise permissible choice that imposes that loss wrong?

We can clarify this question. For any particular way the future might unfold – that is, for any possible future or *world* – we can define a class of *merely possible persons*. Merely possible persons are persons agents *could* but in fact *don't* bring into existence at that world. The choice not to conceive a child, or to conceive a child in vitro but not arrange for the newly created embryo to develop *in utero* into a thinking thing, a *person*, and (by assumption) the choice to have an *early abortion* are all choices that relegate, relative to the world of their performance, *possible* persons to the class of the *merely* possible.

Suppose that agents *do not make* but *could have made* the future unfold for a particular person in such a way that that person comes into existence. And suppose that those same agents *could have made* the future unfold for that person in such a way that that existence itself is *worth having*. Suppose, in other words, agents had the alternative of bringing a particular person into existence *without* dooming that person to either a *wrongful life* – a life *less* than worth living – or a *perfectly neutral* life – a life whose burdens *perfectly counterbalance* whatever it is that is precious in life to the one who lives. And suppose, finally, that we are not in a case where the choice *not* to bring a person into an existence worth having makes things better for anyone at all. This not a case where agents, for example, *must* leave one person

[16] The argument against wrongdoing that is proposed here may sound vaguely like the problem of *collective harm*. The problem of collective harm is urgent. Where, *whatever* I as an agent do, others will suffer by virtue of what still other agents will do, it may seem unclear whether what *I* do can harm those others. And it, accordingly, may seem unclear what my obligations in that context really are. I argue elsewhere that the problem of collective harm does not absolve us from our *participation* in the group act that leads to that harm. I can, as a *participant in a group*, impose harm even if I have not, as an *individual*, imposed harm. The problem of collective harm arises in virtue of the fact that we have overlapping agents: groups of individual agents, overlapping with individual agents. The *group* may have a way to create more wellbeing for a person rather than less when the *individual* does not. Recognizing that fact can help us, I argue elsewhere, to address the problem of collective harm. See Roberts (2006). But the problem presented by late abortion is quite different. There, the question is whether, if everyone else is blithely doing wrong things, why is it wrong for me to do wrong things as well. And the answer is: it just is; it is wrong to do wrong things.

out of existence altogether in order to avoid making things much worse for many others.

Must we then say that the choice not to bring a person into an existence worth having is *wrong*? Must we at least say that agents have a *moral reason* to choose otherwise – or that, *other things being equal*, or *ceteris paribus*, that choice is wrong, or that it is *prima facie* wrong?

After all, there is at least a case to be made – and I think an exactly correct case – that the choice not to bring a new person into an existence worth having is a way of creating less wellbeing for that person when agents could have created more – that that choice makes things worse for a person when agents could have made them better. There is a case to be made, in other words, that the choice not to bring a person into an existence worth having causes that person to incur – though obviously not to *suffer* – a *harm*, or *loss*.

Given, then, that the loss is *incurred*, does it *count*? Does that loss *bear on* the permissibility of the choice that imposes that loss or the alternatives that avoid it? Is that loss *morally significant* for purposes of evaluating that choice? Does it *count against* that choice? Does that loss, in a roundabout way, count *in favor of* the choices that avoid it? If the choice itself is otherwise permissible, *can that loss make that choice wrong*?

1.3.2 Let's first consider the claim that, yes, the loss under scrutiny can make the otherwise permissible choice that imposes that loss wrong. This claim may seem plausible on its face. Some theorists might even consider it obvious that, when we don't have a good reason *not* to bring a person into an existence worth having, we are *obligated* to bring that person into that existence.

Theorists who accept this claim seem to be adopting a radically *inclusive* position in respect of the merely possible. They may be thinking that the *merely possible matter morally*. They may be thinking that we have *obligations in respect of the merely possible*. And they may be concluding that all of the losses incurred by the merely possible, *including* the losses they incur when we choose not to bring them into existence to begin with, are *morally significant* for purposes of evaluating the acts that impose those losses and the alternatives to those acts that avoid those losses.

Now, there seem to be two basic ways of grounding this position. The first rationale asserts that we have obligations in respect of merely possible persons as individuals, one-by-one *for their own sake*. Such obligations can be described as *person-affecting*, or *person-based*, in nature: they are obligations to *affect* (under certain conditions) *particular persons for the better and not for the worse*. Since one way of affecting persons for the better seems to be to bring them into existences worth having in place of none at all, the argument can be made that this is something we are morally required to do – at least when we can do it without making things too much worse for too many others.

To say that we have obligations *in respect of* the merely possible is not necessarily to say that those obligations are *owed to* anyone at all. Still, talk about our having obligations in respect of *merely possible* persons as individuals, one-by-one for their own sake, may seem peculiar. And certainly the *specific* implication that leaving a person out of an existence worth having could ever on its own ground a finding of wrongdoing is ruled out by the narrow person-based principles I have elsewhere proposed.[17]

But even if talk about our having obligations in respect of *merely possible* persons as individuals, one-by-one for their own sake, seems peculiar, the idea that we have such obligations, and that we are accordingly obligated, in many cases, to

[17] The person-based principles I have elsewhere described include the *person-affecting*, or *person-based*, intuition. That core intuition provides just a *necessary*, not a *sufficient*, condition on wrongdoing. The particular version of the person-based intuition I elsewhere adopt – and that is part of Variabilism in combination with an otherwise plausible permissibility theory and specifically OPPP2; see part 2.2.4 below – is the traditional narrow version. According to that principle, an act performed at a world is wrong *only if* that act makes things worse for at least one person *who does or will exist* at that world than things are for that same person at some alternate world accessible to the agents. The act that, performed at a given world, is *bad* must be *bad for* – must, that is, *harm*, or impose a *loss* on – some person who does or will exist at that world. Put in the terms of Variabilism: that act must impose a *morally significant loss* on a person.

Under this narrow version of the person-based intuition, an act that leaves a person out of an existence worth having cannot be declared wrong just on the ground that it imposes a loss on that *merely possible person*. Again, in terms of Variabilism, the loss that act imposes is not itself *morally significant*. If such an act is wrong at all, it is wrong on other grounds.

A "wide" version of the person-based intuition, in contrast, might assert that the loss incurred when a merely possible person is left out of an existence worth having can on its own ground a finding of wrongdoing. This version is represented by Inclusion in combination with OPPP2; see part 2.2.4 below. The wide version of the intuition is not very interesting. (We may even think it misses *the intuition* altogether.) In any case, OPPP2 itself is consistent with both the wide and the narrow version of the person-based intuition.

At the same time, the version of the person-based intuition I describe elsewhere and adopt here (as an implication of Variabilism in combination with OPPP2) is broader in a certain critical respect than what John Broome calls the *Neutrality Intuition*. Broome (2004), p. 143; see also part 2.10 below and Appendix C. On the version I adopt, the necessary condition on wrongdoing is satisfied whenever agents create less wellbeing for an existing or future person when they *could have created more* for that same person. According to this view, when we compare two acts in order to determine their respective *moral betterness*, and ultimately the wrongness of one of those two acts, the loss that counts toward moral *worseness*, and ultimately wrongness – the loss that counts as *morally significant* – can be established, not only by our noting that the one act is worse for an existing or future person than the second act is, but also by our noting that the one act is worse for that person than *some third act*. The Neutrality Intuition operates quite differently, and Broome – rightly, I think – rejects it. The Neutrality Intuition itself, along with the related Prior Existence View that Singer earlier articulated and rejected (Singer 1999, pp. 102ff.), is discussed in part 2.10 below. Broome's more sweeping objection against the person-based approach generally is discussed in Appendix C below.

Another clarification: the person-based approach I have elsewhere described is not equivalent to Exclusion (or Moral Actualism); my principles (barring an occasional lapse; see Roberts (2002), p. 328, where it should have been made explicit that each person who "exists in Y also exists in X") do not leave the merely possible and all their losses out of the picture altogether. They do not, that is, imply that merely possible persons *cannot* be wronged – or, more specifically, that the losses they incur *cannot* make the acts that impose those losses wrong and *cannot* make the acts that avoid those losses right. My principles by their terms apply to all acts performed at all worlds, actual or not. And they are, at the same time, restricted to reflect the point that an act that makes things worse for, for example, *actual* persons is not necessarily wrong in a case where the only way of making things better for those persons would have been to bring still other persons into existences that are less than they might have been. See Roberts (1998), (2002), (2003a), (2003b), (2004), (2006), (2007), (2009).

create more wellbeing for them rather than less, including by way of bringing them into existences worth having, is something we seem able to make sense of.

A second rationale may sound less peculiar but just as effectively ground the idea that we have obligations in respect of the merely possible and that all of the losses they incur have moral significance for purposes of deciding what we ought to do. This second rationale is *impersonal* in nature. It states that a very important moral obligation, and perhaps our only moral obligation, is to *make things better*, not for any person, existing, future or merely possible, but rather *better simpliciter*. We are obligated, in Sidgwick's phrase, to make things better "from the point of view of the universe."[18]

This second rationale, like the first, asserts that we have obligations in respect of the merely possible. But, more clearly than the first, it does not require that we say that those obligations are *owed to* anyone at all. We can be obliged to make certain choices – choices that create additional wellbeing on an aggregate basis, or choices that create additional quantities of the *overall good*, which might itself be an amalgam that includes, in addition to aggregate wellbeing, goods of other sorts as well (such as equality). On this second rationale, we have such obligations not because those choices make things better *for any person*, existing, future or merely possible, but rather because those choices *make things better simpliciter*. We can then say that bringing people into existences worth having is often a way of doing just that.

The claim, then, that the loss incurred by a person we never bring into existence to begin with can make the otherwise permissible choice that imposes that loss wrong is one we can explain. We can say that the merely possible *matter morally*, and that *we have obligations in respect of them*, either one-by-one for their own sake, or on impersonal grounds. And we can say that, accordingly, the losses they incur – *including* the losses they incur when we leave them out of existence altogether – *count against* the otherwise permissible acts that impose those losses. Their losses can make those acts *wrong*.

1.3.3 The other of the two possible answers to our initial question is just that, no, the loss incurred by a merely possible person when we leave that person out of existence altogether *cannot* make the otherwise permissible choice that imposes that loss wrong.

Variabilism enables me to make this point more clearly. Thus: a person who counts as merely possible relative to a world w but who does or will exist relative to a world w' can incur a morally significant loss at w' – a loss, that is, that *bears on* the moral status, not just of acts that *impose that loss at* w', but also of acts that *avoid that loss at* w.

The critical point is that Variabilism, in combination with the otherwise plausible permissibility theory, clearly establishes that a person-based approach is *not* equivalent to Moral Actualism. A person-based approach *need not*, even if some do, put *actual* persons – or even persons who *would be actual* were the particular act under scrutiny itself performed – on a moral pedestal. Among other things, this means that, even though Caspar Hare's arguments against Moral Actualism were not mistaken, his conflation of Moral Actualism with the person-based approach was. Hare (2007), n. 5.

[18] Sidgwick (2006/1884), Preface.

Theorists who make this claim seem to be adopting a radically *exclusive* position in respect of the merely possible. They may be thinking that we have no moral obligations in respect of merely possible persons at all – or, at least, that the merely possible do not matter morally in the same way that you and I do. How could they? They don't, and won't, exist. It might accordingly seem that none of their losses, including the losses they incur when we fail to bring them into existence to begin with, have any moral significance whatsoever.

Suppose that a young woman has chosen non-conception, or conception followed by non-fruition, or early abortion. We agree, of course, that the young woman matters morally. So does her partner and her already-existing and even her future children, and no doubt her grandmother and many others as well. We think their losses are morally significant and must be taken into account in determining whether what the young woman has done is permissible. We think their losses, if any, bear on the permissibility of her choice of early abortion. But we might deny that the losses incurred by merely possible persons – including the loss incurred by the child the woman never brings into existence to begin with – have any moral significance at all.

This is not to deny that there's a *loss*. It's just to say that that loss cannot make the otherwise permissible act that imposes that loss *wrong*.

1.3.4 Our initial question was this: does the loss a merely possible person incurs when we leave that person out of existence altogether count against the choice that imposes that loss? Does such a loss make the otherwise permissible choice that imposes it *wrong*?

We now have two answers to that question – one that claims that there *is* something wrong with the choice not to bring the merely possible person into an existence worth having in place of none at all, and one that claims that there *isn't*. And we can provide rationales for each of those positions. We can adopt Inclusion and say that we *do* have obligations *in respect of* (though not necessarily *owed to*) the merely possible, whether, one-by-one for their own sake or on impersonal grounds, for the sake of the universe, and that their losses accordingly can make the otherwise permissible acts that impose those losses wrong. Or we can adopt Exclusion, and say that we *don't* have obligations in respect of the merely possible and that the losses incurred by the merely possible accordingly *cannot* make the otherwise permissible acts that impose those losses wrong. Under Inclusion, the merely possible *matter morally*, and *so do their losses*, including the loss of never coming into existence at all. And under Exclusion they *don't* matter morally and *neither do their losses* – especially not the loss of never coming into existence at all.

1.3.5 The position that *includes* the merely possible among those who matter morally and deems all their losses to have full moral significance – *Inclusion* – seems at odds with even the most basic conception of procreative privacy. For Inclusion appears to ground a powerful argument against non-conception, against conception followed by non-fruition *and* against early abortion. At least, Inclusion vastly complicates what we think should be a summary moral analysis.

Suppose that a young woman chooses to have an early abortion when she could have instead continued her pregnancy and brought a new person – the person who is relegated by her choice to the class of the merely possible – into an existence worth having. And suppose that Inclusion is true – that merely possible persons matter morally and their losses count. In evaluating her choice, we are then compelled to take into account *not just* the losses incurred by or avoided on behalf of the woman herself and each other person we all agree matters morally – her partner, her already-existing children, her future children, her grandmother, etc. – but also the loss incurred by the merely possible child. And we must recognize that the loss incurred by that child is very great and cuts very deep: the woman's choice of early abortion leaves her possible child *with nothing at all*. In contrast, the woman herself and each other person we agree matters morally may well have already accrued a good deal of wellbeing over their lifetimes and may well reasonably expect to accrue a good deal more whether the early abortion is performed or not. Let's suppose that this is in fact the case. Trading off, then, the various losses against each other – the pregnant woman's, for example, if she proceeds with the pregnancy, against her possible child's, if she does not – we seem compelled to say that the woman's choice of early abortion – or non-conception or conception followed by non-fruition – is *wrong*.

1.3.6 But this result seems completely untenable. We seem here to have generated an *implausibly stringent procreation obligation* – and not just on the part of young women. Surely we must reverse course and say that merely possible persons, after all, *do not matter morally* – that *existing* persons matter morally, and *future* persons matter morally, but the merely possible matter morally *not at all*.

The position that the merely possible are to be *excluded* from those who matter morally and that their losses are, accordingly, devoid of moral significance – *Exclusion* – seems, at first glance, the far more commonsensical position. Isn't it just obvious that what is important is for us to "make people happy" – that "making happy people" cannot really be on our moral to-do list[19]? Surely it is enough that moral law demands that we see to the needs and interests of all those persons who do or ever will exist. But it seems too much – a perversion rather than an expression of moral law – to think that we must also see to the needs and interests of all those many, *many* persons who will never exist at all.

1.3.7 What makes things interesting, however, is that we *can't* reverse course completely without finding ourselves in, if anything, a still *more* untenable position. For, if nothing else, we should be able to agree that distinct persons exist in distinct worlds. *We* need *not* have existed; persons other than ourselves *might* have existed. This means – as we shall see – that it is not going to work to say existing and future persons matter morally but the merely possible matter do not. It is not going to work to say that *we*, at *this* world, matter morally and *they*, at *that other* world, do not.

Now – as we shall also see – Exclusion can itself be developed in either of two ways, *Alpha* and *Beta*. But Exclusion in both its forms collides with the fact that our assessments of what goes on at the distinct worlds and within their distinct

[19] I am paraphrasing Narveson (1976), p. 73.

populations are closely linked to each other. The upshot is that the very same act, under Exclusion Alpha, will be deemed obligatory if unperformed and wrong if performed, a result that unfathomably assigns a moral significance to the fact of performance, and, under Exclusion Beta, will be both permissible and wrong in the very same case – a result that seems clearly inconsistent. To put the point – for now – roughly: the idea that existing and future persons matter morally and the merely possible do not is going to come at the cost of any stable sense of just what we are trying to say when we say that one act is permissible and another is not.

1.3.8 So we have a dilemma: a completely untenable position if we include the merely possible among those persons who matter morally and say that all their losses have moral significance for purposes of determining what we ought to do, and a still more untenable position if we exclude the merely possible, and say instead that none of their losses have moral significance for purposes of determining what we ought to do. Inclusion is untenable, and Exclusion, in both its forms, is, if anything, still more untenable.

1.4 Variabilism as Middle Ground

The idea that Exclusion and Inclusion are our only options – that, if we can't be Exclusionists, we must be Inclusionists – sometimes seems to drive the argument in the minds of some theorists. But it is a false dilemma. And that that's so can be seen when we take care to focus, not on who matters morally and who does not, but rather on what losses matter morally and what losses do not. We can then easily reject Exclusion in both its forms without accepting Inclusion. Instead of saying that *no* losses incurred by any merely possible person have any moral significance at all, or saying that they *all* do, we can say that *some* losses incurred by the merely possible have moral significance and *some* do not. *Exclusion* is a mistake, but so is *Inclusion*.

And we can exactly the same thing about the losses incurred by all of the rest of us as well. We can say that *we all* – existing and future persons, *and* the merely possible – matter morally *and* in exactly the same ways. We all matter morally, but we all matter variably. Now, this might come as a surprise. We might have thought that, having made it into existence, we ourselves are to be accorded some special moral status. But that's not so, according to Variabilism. Some of our losses have moral significance, and some of our losses do not.

Moreover, the variability we see in moral significance is not at all random. There is a distinct pattern. Thus:

> *Variabilism*:
>
> Any loss incurred in any world by any person has moral significance for purposes of evaluating an act that imposes that loss and any of that act's alternatives *if and only if* the person who incurs that loss does or will exist at the world at which that loss is incurred.

On this view, what imbues some losses but not others with moral significance is just *where* that loss is incurred in relation to the person who incurs it. Incurred at world where the person does or will exist, the loss has full moral significance. Incurred anywhere else, the loss has no moral significance at all.

1.5 Variabilism and Abortion

1.5.1 Exclusion and Inclusion are each very extreme positions. As noted above, they accordingly leave a good deal of middle ground between them. They leave room for *Variabilism*. And that room, in turn, opens the door to a certain middle ground on abortion.

The implication from Variabilism that I will argue is important here for dealing with early abortion is just that the loss incurred by a person at a world where that person is never brought into existence to begin with has no moral significance whatsoever. And the implication that is important for dealing with late abortion is – I will argue – just that the loss incurred by a person at a world where that person does or will exist has full moral significance.

1.5.2 Specific permissibility results will need to wait on the otherwise plausible permissibility theory. That theory is spelled out in Chapter 2 below. And, as noted earlier, for purposes here, certain critical assumptions relating to abortion will be adopted. One is that the early abortion is the choice not to bring a person into existence to begin with. Another is that late abortion is the choice to end the life of an already-existing person. We will say more about these assumptions, and others, briefly in part 1.6 below and in more detail in Chapter 5 below.

Still, looking ahead, we can outline the position that Variabilism in combination with the otherwise plausible permissibility theory will take regarding early abortion. *Early abortion* is ordinarily perfectly morally permissible so long as it is what the pregnant woman wants. That is so, since the only person who will typically incur any loss – or at least any loss that is on par with the loss the woman may incur if she continues the pregnancy – as a result of the early abortion is the person the embryo or early fetus will develop into if the pregnancy proceeds and the person who never comes into existence at all (never having had that first thought) if the early abortion is performed. According to Variabilism, that loss, however great or deep, is devoid of any moral significance whatsoever. It thus cannot make the otherwise permissible choice to have the early abortion wrong.

1.5.3 Again looking ahead, we can say that the position that Variabilism in combination with the otherwise plausible permissibility theory will suggest regarding late abortion is this. *Late abortion* is ordinarily wrong. Late abortion ends the life of an already-existing person, a being that has had that first thought and acquired the capacity to survive. It thus causes that person to incur a loss that, according to Variabilism, has *full moral significance*. Moreover, that loss is typically very great and very deep, at least when compared against other losses that might be incurred by other persons if the late abortion is not performed. It, accordingly, cannot be justified under an otherwise plausible permissibility theory. That is so, even in cases in which the woman's own health or even life is at risk or the child faces an existence that is worth having but unavoidably flawed. The exception is the unusual case where the child's life will inevitably be either wrongful or perfectly neutral. In that case, the late abortion imposes *no loss at all* on the child and hence no morally significant loss.

1.5.4 Variabilism thus suggests – for any typical case – a hard line on late abortion and almost no line at all on the early abortion. Early abortion is typically *perfectly* permissible – and a choice we should find permissible on the basis of a *very* summary moral analysis if it is what the pregnant woman herself wants. There is no point in any nuanced, probing, multifarious, dense analysis of just exactly how (A) the loss – if any – the woman may incur if the pregnancy is allowed to continue is to be traded off against (B) the very great and very deep loss her merely possible child will (we agree) incur if the early abortion is performed. In fact it is a symptom of moral confusion, not moral sensitivity, to think that determining the permissibility of the early abortion requires us to carefully balance (A) *any loss at all* (morally significant or not) against (B) a loss that is *completely devoid of moral significance*. Such balancing acts can be sensible *only if* the losses that are to be balanced against each other *both have moral significance*.

The choice of early abortion is thus not ordinarily a particularly hard case. But nor is the choice of late abortion. For there we are dealing with a loss that has *full moral significance*, according to Variabilism, in virtue of the fact that the new person who will incur that loss if the late abortion is performed already has come into existence and acquired the capacity to survive. That is so, by definition of *late abortion*. That loss is, moreover, very great and very deep – at least, compared to the losses others can expect to incur if the late abortion is not performed. Thus, even when the pregnant woman herself has much to lose if she continues the pregnancy, the late abortion typically will be wrong.

Cases involving wrongful life – or even the perfectly neutral life – are the exception.[20] In those cases, the late abortion imposes *no loss at all* on the already-existing person. If the case involves the genuinely wrongful life, the life *less* than worth living, the late abortion *avoids* a loss – and a morally significant loss, at that – on behalf of that person. Such cases may be rare. But they still need to be taken into account.

1.5.5 The construction of a middle ground is sometimes a matter of trying to make things fit together that don't. It's a way of pulling ideas together from both sides of a debate in a way that really just makes things conceptually worse – harder to understand, test and apply, and harder to situate within a broader moral theory.

[20] We all are vulnerable to having our lives made less than worth living, however well-off we might now be or might have been. *Wrongful life* is a more limited concept. The life is *wrongful* when the existence is less than worth having *and* where an existence worth having, or at least a perfectly neutral existence, is not something agents *could* have brought about. In the typical wrongful life case, the possible world where the person born with the serious genetic or chromosomal disorder – so serious that the life is made less than worth living – is given a life that is worth living or at least neutral is *inaccessible* to agents. We do not, that is, yet have the medical technology required to repair or ameliorate the underlying disorder.

But I believe that the two middle grounds I propose here, one on the moral significance of merely possible people and the other on abortion, make things better in those respects rather than worse. We are left with two positions that are both plausible in their own right and fit nicely together. They together enable us to maintain a highly intuitive position regarding abortion without having that position kicked out from under us by an idea about the merely possible that we mistakenly think we are compelled to accept. We can accept the permissibility of early abortion free and clear, in other words, once we reject the idea that the loss incurred by a person when we leave that person out of existence altogether has any moral significance at all.

1.6 Thinking Things, Persons and Abortion

1.6.1 The idea – and an assumption here – that *persons* are the kinds of thing we generally have obligations in respect of and the kinds of thing we often must make things better for *at least* if they exist is not especially contentious. Many of us think that there is at least a sense of the term *person* that implies exactly that moral weight.

Accordingly, other than perhaps Variabilism itself, the most contentious point made so far may just be that a pregnancy does not involve a person until the point in the pregnancy at which *thinking* itself has emerged. The most contentious point is that *persons* are the kinds of things whose coming into existence is *signaled by their thinking*. On this way of looking at things, the embryo and the early fetus are live human organisms. But, until thinking has emerged, there is no person there.

By implication, then, the abortion that destroys a human embryo or early fetus that has not yet had that first thought does not impose a loss on an already-existing person but rather thwarts the coming into existence of a new person.

This account of when something is a person – when a thing is the kind of thing that matters morally – could be called the *Thinking Thing Account*. While I believe that account is plausible and will say some things in its favor in what follows, for purposes here it functions as an *assumption*.

Of course, there are competing views. What we can call the *Human Organism Account*, for example, asserts that human embryos and early human fetuses all count as *persons*. For purposes here, however, that account is, as a matter of assumption, false.

1.6.2 A question, however, immediately arises. Whatever our views about the obligations we have in respect of *merely possible* persons, we can surely agree that ending an *existing* person's life is morally problematic and often wrong. By making the assumption that the human embryo and early fetus are *not* persons – by accepting the Thinking Thing Account as an assumption – I escape the immediate implication from the Human Organism Account that the early abortion ends the life of an existing person and is, accordingly, often wrong. So haven't I begged an important question regarding the moral permissibility of the early abortion?

No. A question for this book is the moral significance of merely possible persons. A second, closely related, question is whether the early abortion is wrong *in virtue of what it does to the merely possible person*. But what is *not* a question for this book is whether the early abortion is wrong *in virtue of what it does to the already-existing human embryo or early human fetus.* I am, accordingly, begging no question of interest here.

For purposes here, then, *if* the early abortion is wrong at all, it is wrong *not* in virtue of what it does to the already-existing human embryo or early human fetus but *rather* in virtue of what it does to the *merely possible person* – the person whose coming into and continuing in existence is prevented by the early abortion.

1.6.3 It's true that the abortion theories I examine in what follows, when put together with certain deontic axioms and other principles we may find hard to reject, build cases against the permissibility of early abortion in many cases in which I will argue in favor of the permissibility of early abortion. At the same time, none of the theories I examine – the theories, that is, from McMahan, Hare, Marquis and Harman – are rooted in the idea that the human embryo or early human fetus – that is, the live human organism that has not yet has its first thought – is a person. Rather, to the extent that those theories put pressure on us to conclude that the early abortion is often wrong or at least morally suspect, they do so on the ground that the human embryo or early fetus can be correlated with a certain *possible person* – a certain person who shall remain *merely possible* if the early abortion is performed.

In other words: according to the theories I will be examining, it is what we do to the *merely possible person*, not what we do to the *human embryo* or *early fetus*, that makes the early abortion wrong in any case in which it *is* wrong.[21] In adopting the Thinking Thing Account, I am accordingly begging no question that is at issue in this book.

1.6.4 While the Thinking Thing Account is an assumption for purposes here, it is worth noting that it is a highly plausible assumption. We need only look at the issues in first-person terms. *I* matter morally. And *I'm* a thinking thing; at least, I *intermittently*[22] think. But I don't think *I* will continue to exist after I have whatever last thought it is I will ever have. It is perfectly possible that *I* will go out of existence even though (however unfortunate for others) my body lives on for awhile, its vital signs strong.

But we should say about the beginning of life what we say about the end: just as *I* may cease to exist even though my body survives, so was there a time when *I* had not yet come into existence even though my body had. The same is true of *you*. We, unlike our bodies, are the kinds of things whose coming into existence and going out of existence are signaled by our *thinking*. And we, moreover, are surely among the kinds of things that *matter morally*, if variably.

---

[21] Thus Earl Conee notes that the following argument can seem sensible to any theorist who thinks that an act's permissibility is determined, at least in part, by the consequences of that act: "Assuming that an early fetus is not a person, consequentialist considerations still argue against the moral permissibility of some early abortions. If an early fetus would grow to become someone who would lead a sufficiently valuable existence, then on objective consequentialist grounds it would be seriously wrong to kill the fetus." See Conee (1999), p. 629.

[22] On the view I describe, once in existence, a person can continue to exist even if that person does not continuously think. *Intermittent* thinking alone is enough to keep one in existence. The person doesn't cease to exist, then, until the conclusion of that last thought – even if their bodies, sustained by natural or artificial means, manage to cling to life far beyond that point. In short, our first thought signals our coming into existence at a world and our last thought signals our going out of existence at that world, but, between these two points, we *exist* whether or not we *continuously* think.

1.6.5 One other point in favor of the Thinking Thing Account can be noted. What it really says is that *thinking* and *mattering morally* – that is, being a *person* – come together. It says that thinking things and thinking things alone are the kinds of things in respect of which we can have moral obligations, whether one-by-one for their own sake, or on impersonal grounds, as a matter of making things better *simpliciter*.

Perhaps what makes a connection between thinking and mattering morally most plausible is just that thinking things are such that it matters *to them* (whether they think so or not) how we treat them. It matters *to them*, that is, whether we create more wellbeing for them rather than less.

Of course, the concept of *thinking* can be naturally analyzed in many different ways. If it is correct that the reason that some individuals matter morally is that it matters *to them* whether we create more wellbeing for them rather than less, then this fact can serve as a guide to just how narrow or broad the concept of thinking we will want to adopt for purposes here needs to be.

Thus, the pain I feel when vaccinated is enough to show that I am a thinking thing for purposes here – and that I say "Not this *again*" enough to show I have continued as the *same* thinking thing from one moment to the next. But individuals do not need a life plan and or any clear concept of themselves as continuing beings to have these kinds of experiences.[23] It can *matter to them* what we do to them, just as it *matters to me* what is done to me. Moreover, we can take this approach without limiting the concept of wellbeing to simple pleasure. Thus, it can matter to the 10-year-old child whether his parents are saving for college or not, and the parents' saving for college may well create more wellbeing for that child rather than less even if the child himself is completely oblivious to the whole thing. We do not need ever to know that wellbeing is being created for us in order for wellbeing to be created for us.

1.6.6 The Thinking Thing Account is, then, though just an assumption, plausible. It is plausible that, in the case of persons, thinking and coming into existence come together. Accordingly, the early abortion – the abortion that takes place during the period the live human organism, whether embryo or early fetus or even scattered object consisting of sperm and egg, has never had that *first thought* that would signal the coming into existence of a new person – is not a choice to *end the life* of an already-existing and otherwise continuing person. Rather, it's the choice *not to bring that person into existence to begin with* – or, if not that, then the choice to end the life of an existing person whose going out of existence was naturally imminent *in any event*.

1.6.7 I argued above that the Thinking Thing Account begs none of the questions I mean to address here. What would beg the question would be to insist that persons

[23] But we need also to make sure that the concept of thinking we make use of here – that we want to connect with being the kind of thing that matters morally – is not *too* broad. See note 3 above and parts 5.2 and 5.3 below. In my household, there has been a good bit of discussion regarding whether insects *think* in the sense that they feel pain. My son thinks they do: how otherwise could they manage not to get themselves killed the moment they are out on their own? I asked my spouse: do you think insects feel pain? His response was: what do you mean by *pain*? I explained: pain in the sense of what I feel when I say "*Ouch, that hurts*!" Do they say: "*Ouch, that hurts*!"? He replied: yes, I think that's *exactly what happens*. They say "*Ouch, that hurts*!"

who remain *merely possible* at a given world (e.g. the *actual* world), though the *kinds* of things that matter morally, in fact *do not matter morally*. It would beg the question to insist at this early stage that how we treat the *merely possible* – whether, that is, we create less wellbeing for them rather than more – has *no moral significance* for purposes of determining whether the choice of early abortion is permissible. And in fact that is nothing we insist on now.

Moreover, it is nothing we will ever insist on at all. Part 1.7 below makes that point. Variabilism thus in fact rejects the idea that the merely possible do not matter morally – that they have *no* moral status whatsoever. Variabilism instead asserts that merely possible persons do matter morally, but that they matter *variably*, with some of their losses being assigned full moral significance and others assigned no moral significance at all.

1.6.8 One other assumption that I make here bears comment. It is a metaphysical assumption, and it asserts that the person, whose coming into existence is signaled by its first thought, and the early human fetus, whose coming into existence is signaled by such things as cell differentiation and functioning biological systems, are numerically distinct. I think that this is a useful and intuitive way of looking at things. It is clarifying to be able to recognize frankly in connection with the end of life that there are some events my body undeniably can live through but *I* can't. And – on the independent assumption that *mattering morally*, that is, being a *person*, and *thinking* come together – it clarifies things, to recognize that the developing biological organism in the form of the early human fetus comes into existence before the person does.

But in this case the assumption we are dealing with is not critical to the debate. Even if the truth is that the early fetus is numerically identical to the later person – even if what I am calling a *person* is really just the *person-phase* of a continuing biological organism – that fact will not change the substance of the moral problems that we are trying to address here or the results we shall reach regarding early and late abortion. Those problems – the problems of the moral significance of merely possible persons and of abortion – would need to be *rearticulated*, but they won't go away. Nor would any such rearticulation decide for us just what the correct resolution of those problems really is.[24]

---

[24] Thus, if an alternate metaphysics turns out to be correct – if, in particular, it turns out that we are each non-thinking embryos and early fetuses for awhile, and then, when thinking emerges, we become persons – then the debate here will need to be rearticulated. Instead of talking about when a person comes into existence, we will talk in terms of when the live organism transitions from its *non-person-phase* to its *person-phase*. And we will need to revise the definition of *early abortion*. It will become, not the choice not to bring a person into existence, but rather the choice to prevent the live organism from transitioning to its "person-phase." And the issue of moral significance of the merely possible will become, not whether we have obligations in respect of merely possible *persons*, but rather whether we have obligations in respect of individuals who have not yet transitioned and never will transition into their respective *person-phases*. But a different metaphysics is not going to make the substantive moral debate go away. See note 143 below. For discussion of the "irrelevance of metaphysics to the moral issue," see Conee (1999), p. 620 and, more generally, pp. 619–646.

1.7 The New Abortion Debate

1.7.1 The claim that thinking signals the coming into existence of a *person* – that until there is thinking there is no person – is, for my purposes, just an assumption. That same claim is the focus of an older, but important, debate on abortion. In that older debate, we would be talking about whether the human embryo or early human fetus is itself a *person* – that is, the kind of thing that matters morally *at least* if it exists; the kind of thing in respect of which we have moral obligations; the kind of thing we normally ought not kill. And we might then extend the debate in a certain way, following Judith Thomson, and ask whether the fact that the human embryo or early fetus is a person necessarily means that abortion is wrong.[25]

The new debate on abortion takes a different tack. It has no need to consider Thomson's claim that the fact that the early embryo or fetus is a person does not necessarily mean that early abortion is wrong. For the new debate can *assume* that the human embryo and early human fetus are not persons – and can accept that the early abortion, if wrong at all, is wrong not in virtue of what it does to the human embryo or early human fetus but rather in virtue of what it does to the person that embryo or fetus might eventually develop into. The new debate thus focuses on how we should think about the moral significance of the *merely possible person* – the person who will never exist at all in the case where the agent chooses early abortion (or non-conception, or conception followed by non-fruition) over conception and fruition.

1.7.2 It is, then, the new debate and not the old debate that I want to take part in here. The following argument reflects that new debate. In fact, in one form or another it dominates that debate. Even theorists who do not want categorically to endorse the conclusion of this argument have felt forced to qualify their own contrary conclusions in ways that leave us wondering whether they really think real-world early abortions can ever be plainly and unconditionally justified at all.

Thus, we might think that:

1. Merely possible persons *matter morally*, in just the way that you and I matter morally.

And:

2. Early abortion often imposes a very great and very deep loss on a merely possible person when the agent could have avoided imposing that loss without imposing too great or too deep a loss on others.

And from (1) and (2) we might conclude that:

3. Early abortion is *often wrong*.

Premise (1) asserts that we have moral obligations in respect of merely possible persons, whether for their own sake or on impersonal grounds, just as we do in respect of each other. The fact that we create less wellbeing for the merely possible

[25] Thomson (1971), pp. 47–66.

when we could have created more without imposing too great or too deep a loss on others, just as when agents create less wellbeing for *us* when they could have created more without imposing too great or too deep a loss on others, is, according to (1), the sort of thing that can make the otherwise permissible act wrong.

Premise (2) asserts that early abortion often imposes a very great and very deep loss on a merely possible person when the agent could have avoided that loss on behalf of that person *without* imposing too great or too deep a loss on anyone else, including the pregnant woman. The difference is that the merely possible person accrues no wellbeing at all – ends up, that is, with the zero wellbeing level implied by never having existed at all – if the early abortion is performed, while the pregnant woman typically has accrued a good deal more wellbeing than that during her lifetime and can expect to accrue a good deal more whether she has the early abortion or not.

Now, premises (1) and (2) both require further discussion, and we will come back to them shortly. But we should begin by taking a quick look at the conclusion – that is, (3). Just how plausible is (3)? Consider the case where the early abortion – the choice *not* to bring a new person into existence – is good for, or at least not bad for, the pregnant woman, her partner, her already-existing and her future offspring, her parents, her grandmother and *each other person who ever does or will exist.* Is it plausible to think that in that sort of case the early abortion is wrong? And if we think that it is wrong – in virtue of what it does to the merely possible person, the person whose coming into existence the early abortion will prevent – then don't we have to say the same thing about non-conception, and about conception followed by non-fruition? Doesn't (3) take us in the direction of an *implausibly stringent procreation obligation*?

Questions like these show that (3) is at least controversial. At the same time, however, the argument from (1) and (2) to (3) may strike us as perfectly valid. Surely, that is, moral law will insist that early abortion – and non-conception, and conception and non-fruition – is often wrong, if it has already been established both that merely possible persons matter morally in just the way you and I do, and that the choice of early abortion imposes a very great and deep loss on one person while saving still others from losses that are, in any ordinary case, relatively minor. Given (1) and (2), can there really be any moral distinction to be drawn between ending the life of an already-existing person and not bringing a new person into existence to begin with?

I want to suggest, however, that the argument is *not* in fact valid. We can accept (1) and (2) and reject (3). We can recognize, that is, that the merely possible matter morally in just the way that you and I do. And, for any case in which the existence itself is worth having, we can accept that not bringing the merely possible person into that existence is to impose a *loss* on that person. But we can at the same time maintain that early abortion – and non-conception, and conception followed by non-fruition – is perfectly permissible in many cases and indeed in practically all the cases when it is what the woman herself wants.

1.7.3 We will return to the validity of the argument below. It may seem that the simpler strategy for avoiding (3) would be to argue that (1) or (2) or both are false. It may seem tempting, for example, to argue that the merely possible do *not* matter morally in *anything like* the way that you and I matter morally, and hence that (1) is false.

That is the approach that Exclusion takes. Exclusion may seem commonsensical. How can a person who *never exists at all* matter morally in *anything like* the way that you or I, or our already-existing or future offspring, or the needy children on the next block or the other side of the planet, or future generations who may find themselves the victims of global warming, matter morally? Since the merely possible do not and never will exist, why should we concern ourselves with their plights, especially in light of the fact that our doing so will often come at a certain cost to persons who do or will exist – persons, that is, who clearly do matter morally?

In the end, however, we are going to be forced to reject Exclusion. Premise (1), understood in a certain way, is going to have to be accepted as an elemental principle, a veritable pillar, of population ethics. To see this – briefly here, and in more detail in Chapter 2 – just consider what happens when we *reject* (1).

Suppose that we say that the merely possible *don't* matter morally, that the losses they incur have no moral significance whatsoever and that, accordingly, their interests are to be *excluded* from our calculations of whether a given act is morally permissible or not. Serious conceptual difficulties immediately arise.

First, we want to be able to say that it *would be wrong* to bring merely possible persons into existence and then, for no reason that is good enough, to torture them. Moreover, the fact that we know that choice is wrong, even if unperformed, tells us, as we deliberate prior to performance, that we ought stay clear of that particular course of action – that we ought not bring the merely possible into existence and then torture them even if doing so would make things somewhat better for some people who do or will exist and worse for none. But the view that merely possible persons do not matter morally – that their losses are devoid of any moral significance – seems to preclude any such analysis. It seems instead – when put together with the otherwise plausible permissibility theory – to lead us inexorably to the result that – for example – the "genocidal adventures of nonactual dictators" are perfectly permissible so long as their victims remain nonactual – that is, merely possible – as well.[26]

Now, one might try to write this result off – though not, I think, very effectively – on the ground that we need not be concerned when a view generates troubling permissibility results in the case of acts that are *never* in fact performed. But there is a second result that is just as troubling and cannot be dismissed in that way. The position that the losses incurred by the merely possible are to be excluded from our calculations of the permissibility of a given act means, not just that our evaluations of our unperformed (nonactual) acts go awry, but also that our evaluations of our performed (actual) acts go awry as well. We will then be saddled with the result that what we have in fact done is wrong when all we have done is – for example – make things somewhat worse for a few people who do or will exist in order to avoid bringing distinct, merely possible persons into existence and then torturing them.

These two results indicate that it is *far* more difficult than one might have supposed simply to deny premise (1).

---

[26] This example is from Caspar Hare. See Hare (2007), p. 503.

1.7.4 Premise (2) as well seems a tempting target. And (2) is certainly controversial. For it supposes that we can assign a loss to a person at a world where that person never exists at all, which in turn seems to require that we can cogently say that an existence worth having is better for that person than never existing at all.

But (2) on closer inspection is hard to refute. To appreciate the case for (2), it is useful to have a bigger picture in mind. For purposes here, *losses* and *wrongs* are two quite distinct sorts of things. Thus, to say that an act (including any omission) imposes a *loss* on a person at a world is just to make a certain *metaphysical* point: that there is some other world accessible to agents such that that same person has additional wellbeing. On this view, we frequently incur losses – but whether those losses ground any particular moral assessment – that, for example, we have been wronged, or that a wrong has been done – is a different matter entirely. If wellbeing is taken away from one person who is already badly off just so that a distinct, already better off person can have still more wellbeing, the loss incurred by the one person may well signify that a wrong has been done. But in many other cases one or the other of two people will be forced to incur a loss yet we will say that no wrong has been done. Why not? Because we think that those losses have been traded off, or distributed between those two people, in a way that moral law itself directs.

Suppose, then, that we agree that an act imposes a loss on a person at a world just when there is some other world accessible to agents such that that same person has additional wellbeing. It is at least plausible that a person accrues no wellbeing at all, positive or negative – no benefits, no burdens, no goods, no bads, no plusses, no minuses – at any world at which that person never exists at all.[27] But at the same time it is clear that in many cases agents *do* have the alternative of bringing a person they leave out of existence at one world into an existence worth having at another world. From these two points, we seem easily able to infer that the act imposes, at the one world, a *loss* on the person it leaves out of existence.

There is nothing in this account that even begins to imply that the person who incurs the loss *exists* in both worlds. A person's existing in each of two worlds might be required in connection with some sorts of comparisons – when, for example, we say that a person is increasing his running speed faster at one world than another. But there is no reason to think that all comparisons will require the person to exist in both worlds. My mother, for example, need not exist in Beijing for it to make sense to say she has no money there – or for it to make sense to say that she has more money in Fort Worth, where she does exist, than she has in Beijing. Similarly, it makes sense to say that a person has more wellbeing at a world where that person has an existence worth having than at a world where that person never exists at all. If to incur a loss is just to have less wellbeing at one world than one has at another as a consequence of what agents have done at each of those worlds, then we have a loss.

We can even explain why the comparison itself is cogent, even though the person at issue exists at one world but not the other. What we are comparing is a *number* that

---

[27] This point is, of course, controversial. For further discussion, see part 2.2.2 below (the loss of never existing) and Roberts (2003a).

we have associated with one world (or place) and one person against a *number* we have associated with another world (or place) and that same person. On the basis of that comparison between *numbers* – all of which, unlike all *people*, exist in all worlds – we say that a loss has been imposed on that person at the one world.

One final point needs to be made about premise (2). (2) itself is at least *plausible enough* that we cannot comfortably let our acceptance of the moral permissibility of early abortion depend on the idea that (2) is false. *If* we think (2) is false, we are speculating. *If* (2) is false, it is false in virtue of some highly technical points having to do with the sorts of references that we can successfully make into worlds where the persons we aim to refer to never exist at all. Such issues would perhaps be worrisome in this context *except* for the fact that the merely possible persons we aim to be referring to clearly do exist in still other worlds. That means that any references into the worlds where those same persons never exist at all can be construed as references into worlds where they do exist. Even then, we obviously often won't be able to assign to those persons – who, say, relative to *this* world never exist at all – *genuine proper names*. Our references to them and our beliefs about them will not be *de re*. But that's often the case when a person has been or is about to be harmed. When shots are fired into a crowd and someone or another as a result is killed, we can judge that a harm, or loss, has been imposed on a person without ever having a clue just *who* has been harmed or incurred the loss. This is what quantifiers are for. Quantification and its related operations will let us say what we need to say.

1.7.5 In the end, then, (1) and (2) both seem perfectly correct. If so, then our only recourse for rejecting (3) is to reject the idea that (1) and (2) together compel us to accept (3).

And that, I believe, is exactly what we can and should do. In this connection, Variabilism is extremely useful. It shows just how we can accept both (1) and (2) but reject (3).

Let's start with (1). According to Variabilism, we each – you, me and the merely possible – matter morally, but we each matter variably. We can have things this way – we can find this middle ground – if we abandon the attempt to say who matters morally and focus instead on just which losses matter morally.

In other words, if our question is: "Does this merely possible *person* matter morally?" we have to say either yes or no, and both answers – the one that includes the merely possible person in the realm of those who matter morally and the one that excludes that person from the realm of those who matter morally – are problematic. If our question instead is: "Which among this merely possible person's *losses* matter morally?" we can say that *some* of this person's losses matter morally and *some* of this *same* person's losses do *not*. And we can say exactly the same thing about you, me and anyone else who does or will exist: some of our losses matter morally and some do not. Thus: the merely possible matter morally just as you and I do in respect of some of the losses they incur, but matter not at all, just as you and I matter not at all, in respect of still others.

What, according to Variabilism, imbues some of a particular person's losses with moral significance but not others – what makes some losses such that we *must* take them into account and other losses such that we must *not* take them into account in calculating the permissibility of a particular act and its alternatives – is just where those losses are incurred in relation to the person who incurs them. Incurred in a circumstance, or possible future or world, where a person does or will exist, a loss will have full moral significance. Incurred anywhere else, a loss will have no moral significance whatsoever, not even the littlest bit, for purposes of evaluating the act that imposes that loss or its alternatives.

It's true that Variabilism humbles some among us in a certain way. If we ever vaguely thought that our own parents or other agents owed it to us to take whatever steps they could to bring us into existence to begin with – that just as they were required to create more wellbeing for us by not dropping us on our heads as infants, they were required to create more wellbeing for us by bringing us into existence – then we shall have to forego that claim. After all, though actual we undoubtedly are, had we never existed at all, the loss we then would have incurred by never having been brought into existence to begin with would have, according to Variabilism, *no moral significance whatsoever*. It would have been perfectly fine, in other words, for us never to have been conceived, born and (one hopes) adored at all.

This is to accept the idea that our coming into existence, while a matter of fabulously good luck, did not, in itself, make things better from a moral point of view. It has been, we can hope, "a blessing" for our parents, but it was never "a commandment."[28] But so what? The procreative entitlement we might vaguely have imagined to be ours in virtue of our status as actual persons against our parents is surely one we can live without.

Variabilism, thus, fully endorses premise (1): merely possible persons matter morally, just as you and I matter morally. We all matter variably – with the losses incurred by any person, merely possible or not, at worlds where that person does or will exist having full moral significance, and losses incurred at worlds where that person never exists having no moral significance at all, for purposes of evaluating both the acts that impose those losses and their alternatives.

1.7.6 Variabilism is, moreover, perfectly consistent with premise (2). In fact, Variabilism presupposes (2) – or at least is a far more plausible theory if (2) is true than if it is not. If we cannot recognize, on the basis of a comparison between the wellbeing one accrues if one never exists and the wellbeing one accrues if one has an existence worth having, that the early abortion imposes a loss on the person it causes never to exist, it is hard to see how we can recognize a loss, on the basis of an analogous comparison, in the context of wrongful life. If leaving someone out of an existence worth having does not impose a loss, neither does, it seems, bringing someone into an existence that is less than worth having. We should thus want to say, both for the case of early abortion – and non-conception, and conception followed by non-fruition – and for the case of wrongful life that agents have imposed a loss on a person in virtue of the fact that they have done less for that person when they could have done more.

1.7.7 And now we come to (3), the conclusion that early abortion is often – or ordinarily – wrong. Not all losses are, according to Variabilism, morally significant.

---

[28] Heyd (2009), p. 16.

Some have moral significance, and some do not. As noted, what determines whether a loss has moral significance, according to Variabilism, is just where that loss is incurred in relation to the person who incurs it. Incurred at a world where a person does or will exist, a loss will have full moral significance. Incurred anywhere else, a loss will have no moral significance whatsoever.

We thus rightfully insist – according to Variabilism – that the loss incurred in the context of wrongful life has full moral significance. For in that case the person who incurs the loss – the person born, for example, with the serious and unavoidable genetic or chromosomal disease – does or will exist at the same world at which the loss is incurred. But we can also insist – again citing Variabilism – that the loss incurred by virtue of the fact that a particular person never exists at all is devoid of moral significance, since that loss is incurred by that person at a world where that person never exists at all.

And thus we can accept (1) and (2) and reject (3). We can say that the person whose coming into existence is thwarted does indeed incur a loss. But we can also say that that loss has no moral significance. And the loss – under any otherwise plausible permissibility theory – that has no moral significance at all cannot count against the act that imposes that loss. A loss that has no moral significance at all cannot make an otherwise permissible act wrong or, in a roundabout way, make an otherwise wrong alternative to that act right.

1.8 Tradeoffs and Abortion

1.8.1 So far, I have only briefly sketched the conclusions I aim to reach regarding early and late abortion – that early abortions are ordinarily permissible, and under very summary analyses, and late abortions are ordinarily very hard to justify. What I have said almost nothing about so far are the myriad *tradeoffs* that can arise in abortion scenarios. There will be more to be said about tradeoffs in Chapter 5 below. But it is important to say just a bit here to avoid creating any deep misconception starting out.

Let's first consider the early abortion – the choice not to bring a person into existence to begin with. Many cases will not be complicated by the fact that a *tradeoff* in favor of one party or the other will need to be made. In many cases, that is, no one will have any stake in how the choice of early abortion is made *other than* the person whose coming into an existence worth having is at issue and, perhaps, the pregnant woman herself. In such cases, Variabilism, in combination with the otherwise plausible permissibility theory, immediately implies that the early abortion is perfectly permissible. The early abortion is permissible because it avoids morally significant losses on behalf of, or at least does not impose any morally significant losses on, *anyone at all*. It's true that the person left out of existence altogether as a result of the early abortion incurs a loss that is very great and very deep. But, according to Variabilism, that particular loss is devoid of moral significance. The otherwise plausible permissibility theory, accordingly, will not take that loss into account in evaluating the early abortion – which will in turn clear the way for the result that the early abortion is itself permissible.

In any such case, the pregnant woman may have the early abortion for a good reason, for a poor reason or for no reason at all. She need not prove or believe that her fetus is defective, or argue that her health is at stake, or claim or provide evidence that she is the victim of rape or incest. She may do just as she pleases, provided that any abortion takes place within that span of time nature allows between the moment of conception and the moment at which the organism growing within her uterus must be recognized as a thinking and otherwise continuing thing.

1.8.2 But in still other cases the losses that the early abortion stands to impose will have full moral significance, according to Variabilism. The woman's partner, for example, and even her *grandmother* may incur just such losses if the early abortion is in fact performed, since those losses will be incurred at worlds where the persons who incur those losses – the partner; the grandmother – do or will exist. In such cases, we do face *tradeoff scenarios*. And we face *morally significant* tradeoff scenarios at that. For the tradeoffs are not just between one loss that has full moral significance and another that has no moral significance at all, but rather between two losses that each have full moral significance.

Such tradeoffs are never completely straightforward. What we are to do about them is a matter that neither Variabilism nor any of its competitors can help us with. They are matters to be addressed, instead, within the scope of the otherwise plausible permissibility theory.

Nonetheless, we can predict that the otherwise plausible permissibility theory in many cases will imply that the right way of making the tradeoff between the woman and her partner, or the woman and her grandmother, will be for the woman to have the early abortion. For one thing, very often the need to make the tradeoff will be illusory. There will be a way to avoid imposing a morally significant loss on anyone at all. There will be a way to accommodate the interests of the woman's partner or her grandmother, short of the woman's continuing the pregnancy she does not want. Moreover, when the conflict is genuine – that is, when the imposition of some morally significant loss or another is *unavoidable* – the pregnant woman will typically face, if not the demonstrably greater loss, then at least the demonstrably greater *risk* of loss. And risk may well be what is pertinent for purposes of assessing permissibility prior to choice, since we will often have scant information regarding actual future loss for any of the parties involved. Here, too, the otherwise plausible permissibility theory may well instruct that the tradeoff is to be made in favor of the pregnant woman.

But in other cases we will reach different conclusions. Thus, we can imagine cases where the loss that the woman will incur is negligible, and the loss her partner, or her grandmother, faces far more serious. We can thus certainly imagine cases in which the otherwise plausible permissibility theory will instruct – as a matter of moral, not positive, law – that the woman ought not have the early abortion.

1.8.3 A different treatment of late abortion is in order. The pregnancy has been allowed to proceed to the point where the fetus has had that first thought that signals a person's coming into existence. And that person has acquired the capacity to survive from one moment to another as the same person. Accordingly, there is no argument to be made that the late abortion will simply cause an existing person to lose a future that that person would have lost in any event. And let's suppose that we are not in a case where that future would be *less* than worth having. We are not, that is, in a case where the late abortion rescues a person from a genuinely wrongful life.

Accordingly, the late abortion will cause the newly existing person to incur a loss. And, since that loss is incurred at a world where that person exists, according to Variabilism, that loss itself must be recognized as having full moral significance.

Of course, the pregnant woman may well incur a morally significant loss if the late abortion is *not* performed. She, after all, exists, too – and her losses count as well, according to Variabilism.

Nonetheless, it seems that, in any ordinary case, the otherwise plausible permissibility theory will instruct that the tradeoff is to be made in favor of the newly existing person rather than the pregnant woman. The morally significant loss that the newly existing person will incur if the late abortion is performed is both very great and very deep. A whole lifetime's worth of wellbeing will have been lost – and it will be lost by a person who has had very little chance to accrue very much wellbeing to begin with. In contrast, the pregnant woman will often already have accrued a good deal of wellbeing, and she may well expect to accrue a good deal more whether she has the late abortion or not.

On these various moral data, it seems that an otherwise plausible permissibility theory will say that the tradeoff between the pregnant woman and the newly existing person must be made in favor of the new person. There will always be the philosopher's case in which the woman must have the late abortion to save the world – or the five human (or non-human; what is important is that they are *persons*) neonates who happen to be cradled alongside the trolley track. There will be the more realistic case in which the "pregnant woman" herself is a child whose own life to date may itself have been barely worth living (or even wrongful, that is, less than worth living) and whose own future may be very bleak. But in any more ordinary case, the otherwise plausible permissibility theory will deem the late abortion wrong.

1.9 Abortion and the Law

Until we have the otherwise plausible permissibility theory firmly in hand, it might seem that we really cannot know what that theory will say about the choice of late abortion in any case where so many other agents are doing things – lawfully, and apparently with the full blessing of society – that are at least as bad and getting away with them.

I suspect that an otherwise plausible permissibility theory will not consider the fact that most other agents are behaving very badly a justification for any one agent's behaving very badly. That everyone else is doing it may explain from a psychological point of view why I've done something wrong. But it seems that, under any otherwise plausible permissibility theory, it won't make the otherwise wrong act morally permissible.

But does it make it *legal*? It could be argued that the bare fact that persons other than pregnant women are legally permitted to do very nearly as they please to many human beings and virtually all non-human animals means that a law requiring pregnant women to adhere to a far more stringent standard violates constitutional equal protection guaranties.[29] Moreover, the fact that it is wrong for the woman to choose the late abortion cannot, on its own, imply that it is permissible for the majority to force the pregnant woman to do the right thing when they themselves in analogous cases are left to decide for themselves. But these are just arguments – and difficult arguments, at that. The better view is probably that the U.S. Constitution – for example – should be understood to protect the woman's right of early abortion but not late abortion. Read in this way, the Constitution would align the cases in just the way Variabilism suggests, recognizing a loss the state has a compelling interest in preventing in the case of late but not in the case of early abortion.

1.10 A Middle Ground on Abortion

The middle ground I propose on abortion thus takes a very hard line on late abortion and almost no line at all on early abortion. Why is *this* middle ground? Is what I am calling *middle ground* in fact going to offend practically *everyone*? I don't think it should. Surely, political conservatives and religious leaders who disdain abortion at any stage of pregnancy for any reason are most concerned to see abortion recognized as impermissible and to see it banned by law late in pregnancy. My proposal reflects that concern. It considers late abortion – not moments before birth but any time after the person – the thinking thing – has come into existence and acquired the capacity to survive – almost always wrong. Neither serious fetal abnormality (barring the case of the genuinely wrongful life), nor maternal health (physical or psychological), nor family circumstances (poverty or resources stretched thin by the needs of existing or future offspring), nor the conditions of conception (rape or incest), nor the woman's own ignorance of the fact that she is pregnant, nor the 16-year-old girl's status as a minor child, would be enough to justify the late abortion.

By the same token, surely those who think abortion is permissible throughout the pregnancy, including those who consider abortion a matter of bodily autonomy, are most concerned to have that fact recognized in connection with early abortion. My proposal reflects that concern as well. Early abortion is almost always perfectly permissible in view of the fact that the single substantial loss that we can expect the abortion to impose in any case in which it is what the woman wants – the loss incurred by the person whose coming into existence is prevented by the early abortion – will be deemed by Variabilism to have no moral significance whatsoever.

[29] At issue here would be, for example, the equal protection clause of the Fourteenth Amendment of the U.S. Constitution. See part 5.7 below.

My proposal thus effectively asks each of the two camps settle for the ground that is most important to it and concede the ground that is most important to the opposition. The discussion of early abortion will no doubt remain fraught – for political or psychological reasons, or because of an evolutionary instinct many men and no doubt some grandmothers have in common that strongly inclines them to find in moral law edicts commanding women to have babies. The discussion of late abortion may remain fraught as well. In recent years, moral philosophers, by virtue of the fact that they aren't quite clear what to say about the moral status of merely possible persons, have only contributed to the confusion. Their theories either weigh against even the earliest abortion or are so subtle we are not quite sure exactly what is being said at all. My goal here is to unwind things just a bit, by first coming to a clearer understanding of the moral significance of merely possible persons and only then beginning to address the problem of abortion.

Chapter 2
The Moral Significance of Merely Possible Persons

2.1 Who Matters Morally?

2.1.1 It is a matter of dispute whether persons who count as *existing or future*, relative to a particular circumstance, or possible future or *world*, have a moral status that *merely possible persons* lack. Are persons who do or will exist the only persons who matter morally? Are they the only persons whose needs and interests we must take into account in calculating what we ought to do? Can only their losses *bear on* – *count against* – the permissibility of acts that impose those losses? Can only their losses, in a roundabout way, *count in favor* of alternative acts that avoid those losses?

Or is the better view that *merely possible persons* – persons who will never exist at all – matter morally too? Must we take their needs and interests into account, alongside our own, in determining what we ought to do? Do losses incurred by merely possible people – like the losses incurred by you and me – *bear on* the permissibility of the acts that impose those losses? Can the losses they incur make the otherwise permissible acts that impose those losses *wrong*? Can avoiding a loss on behalf of the merely possible make otherwise wrong acts *right*?

My first goal in this Chapter 2 is to show that the radically exclusive position regarding the moral significance of merely possible persons – what I will call *Exclusion* – and the radically inclusive position – *Inclusion* – are both completely untenable. My second goal is show that there exists middle ground between these two extreme positions, and that this middle ground, what I will call *Variabilism*, is the far more plausible position.

I then briefly discuss what John Broome calls the *Neutrality Intuition* and what Peter Singer calls the *Prior Existence View*. The *Asymmetry* is a good, final test case for Variabilism, and I conclude this chapter with a note on that puzzle.

2.1.2 The view that we are to *exclude* how merely possible persons are affected by a given act and its alternatives in our calculations of what we ought to do – the view, that is, that the merely possible do not matter morally – is sometimes called

M.A. Roberts, *Abortion and the Moral Significance of Merely Possible Persons: Finding Middle Ground in Hard Cases*, Philosophy and Medicine 107,
DOI 10.1007/978-90-481-3792-3_2, © Springer Science+Business Media B.V. 2010

Moral Actualism.[30] Some theorists might be attracted to *Moral* Actualism because they are attracted to *Modal* Actualism. The thought is that, if there are no nonactual persons, then there is no need for moral law to concern itself with their plights. But it seems that we can reject the metaphysical doctrine while retaining the moral doctrine and vice versa. Moreover, Moral Actualism itself comes in two forms, Strong and Weak.[31] But only one of those forms is strictly *actualist.* Only one form takes the position that persons matter morally *if and only if* they do or will exist at the uniquely *actual* world. The other has us say instead that persons matter morally *if and only if* they would have existed – that is, *would have been "actual"* – had the act under scrutiny been performed.[32]

For those reasons, I will leave the term *Moral Actualism* (Strong and Weak) behind and use simply *Exclusion* (Alpha and Beta) instead.

2.1.3 Independent, then, of our metaphysics, Exclusion may immediately strike us as commonsensical. As between Inclusion and Exclusion, at least, it is Exclusion that seems to have the capacity to recognize the critical moral distinction between "making people happy" and "making happy people."[33]

Inclusion, by comparison, seems utterly fantastic. According to Inclusion, we must *include* how the merely possible are affected, right alongside how we ourselves are affected, in making our calculations of what we ought to do.[34] The problem is that – as we

---

[30] See Josh Parsons, who calls the view *Axiological Actualism*, and Rivka Weinberg (unpublished manuscript). Parsons (2002), pp. 135–147. Jan Narveson, too, is sometimes considered an advocate of Moral Actualism. See Narveson (1976). See also David Heyd (2009).

Caspar Hare suggests that Moral Actualism – Exclusion – is tantamount to the person-based approach, which includes (among other things) the person-based intuition. See Hare (2007). In fact, however, Exclusion is just one way of articulating a person-based approach. Variabilism is an alternate and far more tenable way of articulating both the intuition itself and the approach. See note 17 above and part 2.2.4 below.

[31] Hare uses these terms. See Hare (2007), pp. 502–503.

[32] Many theorists find modal actualism an attractive view. Very stringent forms of that view may seem to compel us to adopt Moral Actualism (Alpha) or to dismiss as nonsense any effort to say that a world at which a person exists can be better (or worse) for that person than a world at which that person never exists at all. But such very stringent approaches seem problematic. A plausible semantics, actualist or not, must make sense of the sentence "JFK could have had another child who was a senator but could have been an astronaut instead." See McMichael (1983). See also Rosen (1990) and Sider (2002).

[33] Narveson (1976), p. 73.

[34] Again, many might object to this way of talking. Surely the merely possible do not in fact *have* interests in the same way you and I might have interests in, for example, having the thermostat turned down. To avoid begging the question in favor Inclusion – or against Inclusion, by making Inclusion sound like the silly idea that we must knit and bake chocolate pies for a person who will never exist at all – it is critical not to read any particular substantive moral position into this way of talking about the merely possible. We can, in other words, talk about the merely possible as "having interests" or "incurring losses" – even if at the end of the day we conclude, with the Exclusionists, that those interests and losses are completely devoid of moral significance or, with the Variabilists, that some of those interests and losses have moral significance but some do not. In short, the purpose of talking this way is to achieve additional clarity, not to beg the question.

shall see – such "voices from another world" are hardly mellifluous.[35] "Bring me into existence," they wail – even when for that to be done *real, live, flesh-and-blood* persons must be made to suffer. But how can my choice to make things better for you be wrong when all I have done to make things better for you is leave someone else out of existence altogether?

2.1.4 Later in this chapter, I detail just how extreme Inclusion in fact is. But as it happens Exclusion is just as extreme. If Inclusion takes the merely possible too seriously, Exclusion – as we shall also see – does not take them nearly seriously enough. The upshot is that Exclusion, when combined with an otherwise plausible permissibility theory, generates many permissibility results that we cannot accept – results that are, in some cases, highly counterintuitive and, in still others, conceptually at odds with one another. And that is so for both Exclusion Alpha and Exclusion Beta.

2.1.5 My guess is that the very real issues that Exclusion cannot seem to avoid, in either its Alpha or its Beta form, have led many theorists to think that Inclusion, however fantastic, *must be correct*.[36] The very aspect of Inclusion that makes it seem

---

[35] See Hare (2007), p. 503.

[36] Inclusion, in some form or another, is widely held. It is reflected in the total and average forms of consequentialism – Totalism and Averagism – as well as any form of Pluralism that considers the maximization of aggregate wellbeing among the goods that are to be taken into account in determining whether one outcome is better than another – and ultimately, perhaps, in determining the permissibility of the act that brings about the outcome rather than the other. (If betterness is not pertinent to permissibility, then I am unclear why we are investigating the former and – accordingly – unclear when a theory of betterness fails.)

One reason that Inclusion is so entrenched at present – though it has long commanded a good deal of interest and respect – perhaps has something to do with the *nonidentity problem*. An otherwise plausible permissibility theory that comports with Inclusion will consider the fact that a person *p* would be better off than a *nonidentical* person *q*, in a case where we cannot bring both *p* and *q* into existence and where, while *q's* existence would be worth having, we cannot provide *q* with any additional wellbeing, to mean that bringing *q* into existence in place of *p* would be wrong. Parfit, most prominently, explores whether the nonidentity problem forces us to adopt an impersonal approach in *Reasons and Persons*. Parfit (1987), pp. 351–379. More recently, Caspar Hare explicitly rejects Exclusion, which he calls *Moral Actualism*, in both its forms. In turning to a "de dicto" solution for purposes of addressing the nonidentity problem, he can be viewed as accepting Inclusion. According to Hare, if conception could have been delayed so that "my third child" – *p* – would have been better off than "my third child" – *q* nonidentical to *p* – in fact is, then the failure to delay conception is wrong. How do we explain why that choice is wrong? It can't be that the failure to delay is *better for q* – that *q* incurs a morally significant loss at, say, w2, where conception is not delayed – since *q* incurs no loss at all at w2; *q*'s wellbeing is there maximized. The wrongness Hare's moral radar is picking up at this juncture would seem to have something to do with the stronger "moral pull" (my term, not his) the merely possible but better off *p* exerts on the agent than does the existing but less well off *q*. See Hare (2007), pp. 511–523. For critical discussion, see Wasserman (2008), pp. 529–535, and Parfit (1987), pp. 359–360.

It seems to me, however, that we should not be too concerned with the nonidentity problem in constructing our moral theory. It is not that I think that the choice of depletion or the risky policy, or the choice to enter into the slave child contract or to take the pleasure pill, is permissible, but rather that it is a fallacy to think that we cannot easily discern harm (in a comparative, intuitive sense of that term) in each of those cases. I thus argue that we can resolve the nonidentity problem without resorting to Inclusion. See part 2.2.5 below and Appendix B.

so counterintuitive on its face – the fact, that is, that it deems all losses of all persons at all worlds to matter morally, independent of who actually exists and independent of the act under scrutiny – is also what enables Inclusion to avoid the serious conceptual difficulties that plague Exclusion.

Now, there's no way around the fact that Inclusion generates highly counterintuitive permissibility results in some cases. But we've just noted that Exclusion, despite its air of commonsense, also generates highly counterintuitive permissibility results in some cases. As we shall see, the fact is that both Inclusion and Exclusion generate results that *only about half-way* conform to our intuitions.

So: neither Inclusion nor Exclusion perfectly conforms to our intuitions. Moreover, Exclusion faces deep conceptual difficulties that Inclusion manages to avoid. Are we, accordingly, compelled to make the leap from Exclusion to Inclusion?

For two reasons, we should first look. (A) There is too much at stake. Before we accept Inclusion, we need first to understand just how much we would be giving up. And (B) even if we agree that Exclusion is false, it is a mistake to think we are *compelled* to accept Inclusion. There are options beyond Exclusion and Inclusion, and we should not let a false dilemma force us to adopt Inclusion.

2.1.6 Let's take these two points in order. (A) Before we adopt Inclusion, we should understand fully what is at stake. Inclusion puts the *merely possible into a wildly implausible competition with those who do or will exist.* If we think the merely possible and *all their losses* matter morally in the way Inclusion suggests, we set up "as sure as fate" a "terrible spat" between merely possible persons on the one hand and existing and future persons on the other.

> The air was littered, an hour or so,
> With bits of gingham and calico …
> … the gingham dog and the calico cat
> Wallowed this way and tumbled that,
> Employing every tooth and claw
> In the awfullest way you ever saw ….[37]

Is this a "fate" that moral law can plausibly be understood to *require*? Can moral law *really not* make a distinction between your diminishing the wellbeing level of your own *existing* dog – tearing it to pieces, bit by bit – and your diminishing the wellbeing level of the *Calico Cat* by way of never bringing that cat into existence to begin with? It is all well and good to leave both parties in the "awfullest way" when they both do, or will, exist, we cannot avoid losses on behalf of either and we have distributed those losses between the two in whatever way it is that moral law requires. But the view that it is perfectly permissible for you to allow your own *flesh and blood, existing dog* to be left in the "awfullest way" so that the *Calico Cat* – or

---

Inclusion may also be at work in McMahan's own analysis of the Asymmetry and his treatment of abortion. See McMahan (2009), and (2006), pp. 625–655. I describe the Variabilist account of the Asymmetry in part 2.11 below and McMahan's treatment of abortion in Chapter 3 below. See also McMahan (2002), pp. 627 and 639–642.

[37] Field (1912).

the Cat Who Went to Heaven or Thomasina or any other *merely possible* cat – can be rescued from the *vast repository of individuals who will never exist at all* seems badly mistaken.

I believe that we in the end will be convinced that Exclusion in both its forms is fatally defective. But that fact is not going to make Inclusion seem *any less fantastic* to us than it does starting out. We are never going to say: Oh, *now* I see why Inclusion must be right. The one distinction that Inclusion insists we set aside is always going to seem to us one that no sound moral analysis can conceivably set aside: that one act imposes a loss on a *real, live, flesh and blood, sentient being* and the other a loss on, well, *nothing* that does or will *ever exist at all*. There *just is* an important moral distinction to be made between "making people happy" and "making happy people."[38] In a way that can only be described as axiomatic, your actual dog *must* come before your merely possible cat.

The deficiencies in Inclusion are thus deep enough that rejecting Exclusion in favor of Inclusion does *not get us out of a bind*. It just *tightens the bind we are already in*. The purpose of this Chapter 2 is to *get us out*.

2.1.7 The second reason that any leap from Exclusion to Inclusion would be premature is that (B) there is a plausible middle ground between Exclusion and Inclusion. According to that middle ground, the merely possible do matter morally, just as you and I do, but they matter *variably* – and so do we. This middle ground could be called *Variable Exclusion* or – just as well – *Variable Inclusion*. I will settle for *Variabilism* here.

The strategy behind Variabilism is to identify, not which *people* matter morally and which do not, but rather which *losses* matter morally and which do not. We can then say that, for any person, whether existing, future or merely possible, some of the losses that person incurs matter morally and others do not.

As the cases will show, there is a pattern in when losses have moral significance and when they don't. But that pattern has nothing to do with *who* happens to incur the loss (whether, that is, the person incurring the loss *actually* exists, or *would* exist if the act under scrutiny were performed, or etc.). Instead, what imbues some losses but not others with moral significance – what makes some losses such that, in calculating what we ought to do, we *must* take them into account, and makes other losses such that we *must not* take them into account – is just *where* those losses are incurred in relation to the person who incurs them. Incurred in a circumstance, or possible future or *world*, where a person does or will exist, a loss will have *full moral significance*. Incurred anywhere else, a loss will have *no moral significance whatsoever*, not even the littlest bit.

The availability of Variabilism thus shows that, if, having established that Exclusion is false, we *do* infer that Inclusion is true – if we do make the leap from Exclusion to Inclusion – it isn't because we *must* but instead only because we mistakenly *think* that we must.

---

[38] Narveson (1976), p. 73.

2.2 Preliminaries – A Maximizing Account of Loss; the Loss of Never Existing; the Loss of Death; the Otherwise Plausible Permissibility Theory

2.2.1 *Loss*. Some preliminaries are in order. For purposes here, I adopt a *maximizing* account of harm, or *loss*. I believe that my account of *loss* captures one ordinary concept of *loss* (or *harm*) that we have. For purposes here, however, the important thing is that the term *loss* is shorthand for a certain *betterness* relation. Thus: to say a person incurs a *loss* at one world under one act is just to say that an alternative act performed at an alternative world is *better for* that person than is the one act performed at the one world.

In other words: a person incurs a *loss* whenever agents (by act or omission) create less wellbeing for that person when agents *could have created more wellbeing* for that very *same person*. That is so, even if it is also true that those same agents *would* never in a million years have been willing to create that additional wellbeing for that person.

One feature of the maximizing account of loss will be especially important going forward – in particular, when we begin to compare an act a performed at a world w against an act a' performed at a world w' for purposes of determining the permissibility of act a. On my account, a person can incur a loss at w *not just* in virtue of how that person fares at w relative to how that same person fares *at* w', *but also* in virtue of how that person fares *at some third world* w'' under some alternative act a''. This feature will mean that the loss incurred under a' at w' *relative to* w'' – if that loss is deemed morally significant – can easily count against a', notwithstanding the fact that the person incurs no loss at all at w' *relative to* w (i.e. is no worse off at w' than that same person is at w).

The maximizing account of loss is inherently *comparative*. Incurring a loss is a matter of being made worse off. And we determine whether a person is made worse off by comparing how well off that person is in one circumstance, or possible future or world, with how well off that same person is in another.

Temporal and *counterfactual* (or *but for*) accounts of harm, or loss, also operate by comparing wellbeing levels for the same person from one circumstance to another. But such accounts notoriously undercount loss. The temporal account considers worsening over time to be a necessary condition for loss. But clearly a person can be made better off, or at least not worse off, from one moment to the next and *still incur a loss*.[39] The reason we think that the surgical team who has done a sloppy job restoring your vision – you now can see, but not well – has caused you to incur a *loss* is that we understand that the team *could have done more for you* than it has.[40] Whether the team was *obligated*, morally or legally, to have acted with greater care on your behalf – whether, that is, you have been *wronged* – is a different question. That you are better off does not mean you have not been harmed or that you have not incurred a loss.

According to the counterfactual account, an act imposes a loss on a person *only if*, had agents not performed that act (*but for* that act), that person *would have been*

[39] Feinberg (1988).

[40] This example is based on Hanser (2009).

better off. But, clearly, I have caused you to incur a loss when I shoot you in the arm even if I was so angry that, had I not shot you in the arm, I *would have* shot you in the heart. In a way, I have *saved your life* by shooting you in the arm. But still you incur a *loss*. According to the maximizing account of loss I propose here, it is a simple thing to explain why that is so: I *could have* refrained from shooting you altogether; I *could have* done more for you than shoot you in the arm. The difference between what I did and what I could have done counts as a *loss*.

The maximizing account recognizes, in other words, that a person can (A) incur no loss at all between one world and a second world – for example, the world we count, for purposes of evaluating the counterfactual, as the "closest" world at which the act under scrutiny is not performed – but (B) still incur a loss between the one world and some third – a world agents *could* have brought about, independent of what they *would* have done; a world that was *accessible* to agents at the time just prior to performance.[41]

That the maximizing account of loss is broader than some of its competitors does not mean that we necessarily end up with an unduly broad concept of *wrongdoing*. After all, once we identify loss, we must then determine whether that particular loss is *morally significant*. If we determine that it is, we must then appeal to the otherwise plausible permissibility theory to determine whether a *wrong* has been done.

At the same time, the account of loss that I propose is narrower in certain respects than some other accounts. For example, the fact that it is inherently *comparative* narrows its scope. On my account, a person incurs a loss only if that person is worse off than he or she is relative to some baseline.[42] In contrast, a *non-comparative* approach defines loss by reference to a list of specific bad effects (e.g. substantial pain or impairment) or a threshold (e.g. a minimally decent wellbeing level).[43] But it is unclear to me that a person whose wellbeing has been maximized may also have incurred a *loss*. It is unclear to me that the vaccination – or the kidney transplant – that imposes pain has also imposed a *loss* or a *harm* in any sense that is relevant to the issue of wrongdoing. A maximizing account of loss easily explains why the patient incurs no loss in such cases: when wellbeing has been maximized, there is no *loss*, even if there is pain.

On my account of loss, determining whether an act imposes a loss when performed at a given world obviously supposes that any burdens imposed by that act are to be netted against the wellbeing that is created for that person by that act. This point is somewhat controversial. Where, however, we limit our focus to what a given act does to a given person at a given world, netting seems in order. It is netting

---

[41] I take it for granted that a person's incurring a *loss* at a world requires a cross-world *identity* but does *not* require that person's having an alternate better off "self," in Velleman's sense, at an alternate world (other than to the extent that there being a "self" there coincides with the person who incurs the loss here being *identical* to a better off person there). I thus set Velleman's skeptical arguments aside for purposes here. But see Velleman (2008).

[42] Feinberg (1988).

[43] See Harman (2009), p. 139, and Harman (2000), p. 93, for discussion of a list approach to harm. See also Shiffrin (1999), p. 123. For discussion of the relevance of the minimally decent life, see Steinbock (2009), p. 164.

burdens against benefits that tells us, for example, that the vaccination, despite the pain, does not impose a loss.[44]

In still another way, the account of loss that I adopt here is narrow. On my account, *loss* does not include damage inflicted by nature unless agents could have interceded to avoid that damage. I think there is a familiar sense of the term *loss* according to which some of the deepest losses we will ever incur are thanks to nature. So connecting loss with whether agents could have done more for a person than they have done may seem artificial. On the other hand, there is no reason to think that my account of *man-made* loss cannot coexist with a separate account of *natural* loss. But we have no reason for undertaking that latter account here.

2.2.2 *The Loss of Never Existing.* I accept for purposes here that we can cogently compare a person's wellbeing level at a world where that person does or will exist against a person's wellbeing level at a world where that same person never exists.[45] Thus, as noted earlier, it is at least plausible that a person accrues *no wellbeing at all* – no benefits, no burdens, no goods, no "bads" – at any world at which that person never exists. We can say that persons who never exist at a world have a zero wellbeing level at that world.

We can then say that an agent's leaving a person out of existence altogether at a world often creates less wellbeing for that person when the agent could have created more wellbeing for that same person by bringing that person into an existence worth having. It often makes things worse for that person when things might have been made better for that same person. Accordingly, leaving a person out of existence altogether at a world often causes that person to incur a *loss* at that world.

Similarly, bringing a person into existence often creates more wellbeing for that person when agents could have created less wellbeing for that person by leaving

---

[44] See Benatar (2006), pp. 18–22, for the argument that for purposes of determining harm burdens are not to be netted against benefits.

[45] I have elsewhere argued that comparisons between a person's wellbeing level at a world where that person exists and a person's wellbeing level at a world where that person never exists are cogent. See Roberts (2003a). Since we are here investigating (among others) Inclusion, according to which the merely possible *do* matter morally and the losses they incur when we leave them out of existence altogether (or bring them into an existence that is *less* than worth having) *do* bear on the permissibility of the acts we perform, it can only add clarity to talk openly about such losses. Assigning a positive number to the existence worth having and a negative number to the existence less than worth having (even if those numbers have their ordinal value only) and zero to the state of never having existed at all is also useful. See too Nils Holtug (2001), pp. 361–384, and Fred Feldman (1991), p. 210 ("I stipulate that if s fails to exist at w, then V(s, w) = 0."). But see also Krister Bykvist (2007), pp. 339–345; John Broome (1999), p. 168; McMahan (1999), p. 168; Clark Wolf (1997), pp. 108–109; David Heyd (1992), pp. 30–33; Cynthia Cohen (1996), p. 22; and Partha Dasgupta (1995), p. 383 ("It makes no sense to attribute a degree of well-being, low or high or nil, to the 'state of not being born.' Non-existence is like nothing for us, not even a very long night, because there is no *us* to imagine upon. One can't be asked what it would be like to experience one's own nonexistence, for there is no *subject* of experience in non-existence."). More recently, Broome, citing Wlodek Rabinowicz, has suggested that, even if it is not cogent to say that a person is better (or worse) off "in" a world in which that person never exists, it may still be "analytically possible" to say that a world in which a person never exists is better (or worse) "for" that person. Broome (2004), p. 63. See also note 32 above.

that person out of existence altogether. Bringing a person into an existence worth having thus *avoids a loss* on behalf of that person.

But we must consider the unusual cases as well. We must recognize, that is, that leaving a person out of existence altogether sometimes creates more wellbeing for that person than bringing that person into a genuinely *wrongful life* – an existence that is *less* than worth having – does. In such cases, leaving the person out of existence altogether is what *avoids a loss* on behalf of that person. And bringing the person into existence is what *imposes a loss* on that person.

An implication of this way of valuing the plight of never having existed at all is that a person may incur distinct losses at a given world at which that person never exists relative to the distinct alternative existences he or she might have been brought into. *Anyone's* life can be made less than worth living; *wrongful life* is the case where that outcome cannot be avoided if one is to exist at all.

2.2.3 *The Loss of Death.* It is plausible that, once someone dies, we can no longer act in a way that makes things better or worse for that person. It is plausible that additional wellbeing or illbeing cannot continue to accrue to a person *after* that person ceases to exist. If that is correct, then it is misleading to talk as though agents can perform acts that cause persons who once existed but no longer exist to incur losses. In particular, it is misleading to suggest that the losses they incur when still alive can be made up for by post-existence gains. Accordingly, for purposes here, I avoid any reference to acts that impose losses on persons who *have existed.* The standard reference, instead, is to losses that are incurred by a person who *does or will exist* at the world at which the loss is incurred.

Of course, it would be a problem if an account formulated in this way meant that death was not itself a loss. But it doesn't. Even given that the incurrence of a post-existence loss is not possible, that *last moment of existence* – like any other – is one during which losses *can* be incurred. That the *effect* of a particular act – the effect being that the person *no longer exists* – goes *beyond* that last moment of existence *does not mean* that that person does not *incur* the loss at that last moment. That the *effect* of death materializes only *after* the person goes out of existence does not mean that the *loss* is not incurred *before* that time.[46]

[46] In other words, we thus can discern a loss, when (i) agents act at t1 in a way that makes t2 the last moment of a person's existence and that accordingly generates no future at all for that person at t3 and (ii) those same agents could have instead acted at t1 in a way that would have meant that t2 is *not* the last moment of a person's existence and that would accordingly have generated a future worth having for that person at t3. They have acted in a way, at t1, that imposes on that person – causes, that is, that person to incur – a loss at t2 – a loss that itself is incurred at t2 but doesn't materialize until t3. It isn't, in other words, the *death* that imposes the loss. It's the *act* that imposes the loss – a loss that may be incurred, for example, during that last moment of existence and just prior to death.

By implication, I am here accepting a comparative account of the loss of death. Feldman also describes a comparative account of the loss of death, but the comparison he would make is between the future as it in fact unfolds and the future that unfolds at the nearest world where the person does not die (the future that would have unfolded had the person not died). Feldman (1992), p. 216–217. See also Feldman (1994). But see part 2.2.1 above (discussion of difficulties inherent in counterfactual account of loss).

This means that the death of someone else is not just a loss *we* incur. It's a loss *they* incur as well. They just don't *experience* it, and we do.

But the loss itself is important, not just the experience of the loss. The experience of the loss should be understood as a separate loss. We aren't just grieving for ourselves when someone dies. We aren't just grieving for ourselves and a loss we have incurred. We are also grieving for the person who has died. We are grieving, that is, for the fact that that person has incurred a loss.

2.2.4 *The Otherwise Plausible Permissibility Theory.* The competing rules I examine here are not *moral theories* in a traditional sense. They don't, on their own, decide just which acts are permissible and which are not. Instead, they decide just which *losses* have moral significance and which do not. Such rules – *loss rules* – generate different, and competing, collections of *moral data* on the basis of which an independent and otherwise plausible permissibility theory can then generate different, and competing, permissibility results.

Clearly, if the testing that we do of Exclusion Alpha against Exclusion Beta, Exclusion Beta against Inclusion, Inclusion against Variabilism, and so on, is to be meaningful, the theory that we rely on to generate the permissibility results we use for that purpose must be *otherwise plausible*. The theory must be plausible other than in respect of the losses deemed morally significant by whatever loss rule it is we happen to be testing. Without that assurance, we won't have a clue whether any of the many transparently false results we obtain in what follows are to be ascribed to the loss rule we are then testing or to the permissibility theory.

We thus will need to say something about the content of any such *otherwise plausible permissibility theory*. We won't agree, starting out, just which losses are morally significant and which are not. But setting that issue aside, we will find it easy to agree, I think, on a handful of principles that simply *connect* (A) the moral data they are given by whatever loss rule we happen to be testing, with (B) specific permissibility results, for some range of cases that is of interest to us.

And a handful of principles is all we need. For purposes here, we don't need a *complete* permissibility theory. A rudimentary theory will do. Our needs are modest since the test cases will be simple.

We can then begin with a Pareto principle of sorts.

Otherwise Plausible Permissibility Principle 1 (*OPPP1*):
An act a performed at a world w is wrong *if*

(i) Act a imposes a morally significant loss on a person p at w;
(ii) An act a' at an accessible world w' performed in lieu of a at w is better for p at w' than a is for p at w; and
(iii) Either (a) a' imposes no morally significant loss on any person q at w' other than perhaps p, or (b) if a' imposes a morally significant loss on a person q at w', then a imposes a morally significant loss on q at w, and a' is at least as good for q at w' as a is for q at w.

The Pareto idea – roughly put – is that if agents can make things better for at least one person and worse for none and fail to do so, then what they have done is wrong. OPPP1 restates that idea so that certain critical questions are explicitly left open.

OPPP1 doesn't leave *every* question open. The maximizing account of loss I have adopted here is meant to be incorporated into OPPP1. Thus, for purposes here, it is settled that a loss is incurred by a person when that person is left out of an existence worth having. What clause (i) of OPPP1 leaves open is the moral significance of that loss.

That OPPP1 incorporates the maximizing account of loss settles still another important point. Clause (iii) is potentially failed when q incurs a loss at w'. What is important for purposes here is that loss itself is equally well established by comparing how well off q is in w' against how well off q is in w *and* by comparing how well off q is in w' against how well of q is in some third world w''.

Finally, just as a loss is imposed when agents leave a person out of an existence worth having, bringing a person into an existence worth having can be a way of making things better for that person for purposes of clause (ii).[47]

OPPP1 thus takes as settled certain issues about loss and what counts as making things better for a person. But OPPP1 leaves open whether those losses count as *morally significant losses*. It leaves open whether (A) the loss p incurs at w in the case where p never exists at w and (B) the loss q incurs at w' in the case where q never exists at w' count as *morally significant* losses. Thus, if leaving p out of an existence worth having at w is a way of imposing a morally significant loss on p at w, then OPPP1 may on that basis (and depending on other facts) imply that a is wrong – that it is wrong *not* to bring p into existence. But if it doesn't, OPPP1 will have no application at all. Some of the loss rules we examine in what follows take one position on that issue; some the other. But that issue is part of what is at stake here. So the otherwise plausible permissibility theory – the theory that determines permissibility – must, if it is to be the sort of widely accepted and uncontroversial theory that we can use for purposes of testing, avoid resolving that particular issue in advance.

What distinguishes OPPP1 from some rough formulations of the Pareto idea is this: when we compare an act performed at one world against an act performed at another world for purposes of evaluating the one act, we do not just ask whether things have been made worse for any person at that other world than they are at the one. We ask a broader question: whether a person has been made to incur a morally significant loss at that other world. And that is a matter *not just* of how that person fares at the other world relative to the one, *but also* of how that person fares at any *third* world relative to the one.

Whether any such loss itself has moral significance is something OPPP1 leaves open. Specifically, clause (iii) of OPPP1 leaves open the issue of whether the loss q incurs at w' *compared to how well q fares at some third world w''* counts as a morally significant loss. Or do we say instead that q's loss at w' counts as morally significant, for purposes of evaluating a, *only if* q incurs a loss at w' *compared to how q fares at w*?

---

[47] See parts 2.2.1 and 2.2.2 above.

This issue is potentially important in any additional person context. For, if *q* doesn't exist at all at *w*, then *q* will not incur a loss at *w′ compared to how well q fares at w* so long as *q* has an existence worth having at *w′*. If *that* is the clause (iii) test, then it is easily satisfied. But there is nothing in OPPP1 that indicates that that narrow reading is in order. Consistent with OPPP1, clause (iii) may be far more stringent. That *q* incurs no loss at *w′* compared to how *q* fares at *w* doesn't at all imply that *q* incurs *no morally significant loss at w′ at all.*

And it is critical that OPPP1 avoid taking any position on that issue. Whether the loss *q* incurs at *w′* relative to some third world *w″* counts as morally significant is part of what is at stake here. Some of the loss rules that we will examine will say that it is, and some that it isn't. It is, accordingly, critical that that issue remain open in the context of the otherwise plausible permissibility theory.

Still another question OPPP1 leaves open is its relation to the person-affecting, or person-based, intuition. OPPP1 provides only a sufficient, and not a necessary, condition for wrongdoing. That means that OPPP1 leaves open the question of whether *a* is wrong in a case where agents *do not have the alternative of making things better for the identical person p* who incurs a morally significant loss at w. Consistent with OPPP1, agents have done something wrong – not by implication from OPPP1, but rather by implication from some other sufficient condition on wrongdoing – when they (1) bring *p* into an existence that is unavoidably flawed but worth having and (2) had the alternative of bringing a better off but *nonidentical* person *q* into existence in place of p.

In other words: OPPP1 does not imply the person-based intuition, *narrowly construed.*[48] And that is a good thing, since the truth of the person-based intuition, narrowly construed, is also part of what is at stake here. The loss rules we examine in follows, in other words, take very different positions on just that issue. So it is an issue our otherwise plausible permissibility theory must itself leave open.

In a moment, we will turn to a quite different necessary condition on wrongdoing, a form of the person-based intuition that we *can* include in the otherwise plausible permissibility theory by the virtue of the fact that it leaves open whether it is to be narrowly or more broadly construed. But first there is more to be said about OPPP1.

The skeletal OPPP1 is just one Pareto principle. We examine two others in Chapter 3. The first is the nonstandard Pareto principle sometimes called *Pareto Plus.*[49] The rough idea there is that it is wrong not to bring a person into an existence worth having when agents have the alternative of doing that without making things worse for any person. Pareto Plus is, however, obviously too controversial to be included here as part of an otherwise plausible permissibility theory. Its truth, too, is at stake; and we accordingly are not going to assume its truth starting out. To put the point another way: Pareto Plus is *what we get* when we combine Inclusion with the otherwise plausible permissibility theory (which itself includes OPPP1).

---

[48] See note 17 above (relation between person-based approach and Moral Actualism).

[49] Dasgupta (1995), pp. 382–83. The principle Dasgupta calls *Pareto Plus* is stronger than what I call *Pareto Plus* here. See note 142 below.

The second Pareto principle that we will examine in Chapter 3 is more standard. It is limited to the case where the choices to be compared involve identical populations – when they are, that is, "same-people" choices. OPPP1, in contrast, is not limited in that way. Accordingly, in cases in which leaving one person out of existence makes things better for still others, OPPP1 may *require* agents *not* to bring that person into existence to begin with, depending on the loss rule we in the end adopt. But that result is nothing we are going to build into the otherwise plausible permissibility theory itself. Whether or not that result is correct is part of what is at stake here. It is what we get – and *Pareto Minus* would be good name for it – when we combine Exclusion, or Variabilism, with OPPP1.

A result we will be able to obtain from OPPP1 under some, but not all, of the loss rules we examine in what follows involves the case where the life is both unavoidably less than worth living and actual. The life is, that is, *genuinely wrongful* and the person who incurs it is someone who exists at what is uniquely the *actual* world. The maximizing account of loss immediately implies that that person incurs a loss. Inclusion, Exclusion and Variabilism all agree that that loss is morally significant. Where leaving that person out of existence altogether would not have created a morally significant loss for anyone at all, OPPP1 easily implies that bringing that person into existence is wrong.

A second component of the otherwise plausible permissibility theory is a version of the person-based intuition. It provides not a sufficient but rather a *necessary* condition on wrongdoing – by implication, a *sufficient* condition for permissibility.

Otherwise Plausible Permissibility Principle 2 (OPPP2):

> An act *a* performed at a world *w* is wrong *only if* act *a* imposes a morally significant loss on a person *p* at *w*.

Wrongdoing requires, in other words, that someone or another – some existing person or some future person or, consistent with OPPP2, some *merely possible* person – incur a morally significant loss.

Narrowly construed, OPPP2 will identify quite a range of acts as permissible. But it's nonetheless suitable for inclusion in the otherwise plausible permissibility theory because it is consistent with both a *narrow* form and an alternate *very wide* form of the person-based intuition.[50] OPPP2 thus should not be considered *at all* controversial for purposes here. After all, the class of persons we quantify over in OPPP2 *includes* the merely possible. OPPP2 thus leaves open the question whether its necessary condition is satisfied when the loss that is incurred is incurred by a person who never exists at all. As long as the loss is morally significant, the necessary condition will be satisfied. Some of the loss rules we will be examining say that it is, and others that it isn't.

Neither of the two principles we have included in the otherwise plausible permissibility theory so far addresses overt tradeoff scenarios. In part, that is so because the

---

[50] This latter way of construing OPPP2 may be equivalent to what Parfit calls the "wide person-affecting approach." Parfit (1987), pp. 396–401.

whole topic of how tradeoffs are to be resolved is inherently controversial. Nonetheless, the following principle surely would command wide agreement. The idea behind this principle is that an act is permissible if each alternative act involves a *merely reversing change*.[51] Thus:

Otherwise Plausible Permissibility Principle 3 (OPPP3):
An act *a* performed at a world *w* is permissible *if*

(i) Act *a* imposes a morally significant loss on a person *p* at *w*;
(ii) Each alternative to *a* – that is, each *a′* performed at an accessible world *w′* in lieu of *a* performed at *w* – imposes a morally significant loss on a person *q* (who may but need not be identical to *p*);
(iii) *a* performed at *w* is at least as good for *p* as *a′* is for *q* at *w′*; and
(iv) Either (a) *a* imposes no morally significant loss on any person *r* at *w* other than *p*, or (b) if *a* imposes a morally significant loss on any person *r* at *w* other than *p*, then *a′* imposes a morally significant loss on *r* at *w′* and *a* performed at *w* is at least as good for *r* as *a′* is for *r* at *w′*.

OPPP3 proposes a sufficient condition on permissibility. Suppose that agents have *just two* alternatives *a* and *a′*, and the loss *q* incurs under a′is at least as great and deep as the loss *p* incurs under *a*, and the acts otherwise impose no morally significant losses at all. OPPP3 then implies that *a* is permissible. Moreover, when the two acts both impose morally significant losses, and those losses are themselves equally great and equally deep, then *it doesn't matter* which act agents perform. *Both* acts are permissible; the distinctions between them are, in effect, deemed immaterial.

Clause (i) limits the principle to the case where the act imposes a morally significant loss. If the act under scrutiny imposes no morally significant loss on anyone at all, then we need not bother with OPPP3. We can just appeal to OPPP2 directly to determine that that act is permissible.

OPPP3, like the other principles that qualify for the otherwise plausible permissibility theory, leaves some important issues open. Thus, clause (i) leaves open whether *p* must exist at *w* to incur a morally significant loss at w. Similarly, clause (ii) leaves open whether *q* must exist at *w′* to incur a morally significant loss at *w′*. Some of the loss rules we examine here say that that loss is morally significant even if *q* does not exist at *w′* and some say that it's not. If the loss of having been left out of existence altogether *does* count as morally significant, then two acts in some cases will be considered a *wash* even though one imposes a loss on an existing person and the other on a person who never exists at all. If that loss does not count as morally significant, OPPP3 will have no application at all.

One last point. The rationale for clause (iv) is just to insure that act *a* is not flowed in some way that act *a′* is not and that has nothing to do with *p* or *q*. Clause (iv) thus insures that we will not conclude, on the basis of what *a* does to still

[51] See generally Peter Vallentyne (2000), pp. 1–19.

others, that a is wrong. The two acts are, in other words, otherwise morally indistinguishable – there is otherwise *nothing about the alternative act* that might make us think that the one act is itself wrong.

A final principle declares an act wrong if it imposes a morally significant loss on one person and agents could have avoided that loss on behalf of that person by imposing a less severe morally significant loss on at most one other person. This principle, like OPPP3, addresses a handful of very simple tradeoff scenarios.

Otherwise Plausible Permissibility Principle 4 (*OPPP4*):
An act a performed at a world w is wrong *if*

(i) Act a imposes a morally significant loss on a person p at w;
(ii) An act a' performed at an accessible world w' in lieu of a performed at w is better for p at w' than a is for p at w;
(iii) If a′ performed at w' imposes a morally significant loss on any person q at w', then that morally significant loss is not as great and not as deep as the morally significant loss that a performed at w imposes on p; and
(iv) Either (a) at most one person – q – incurs a morally significant loss under a' at w' or (b) if a' imposes a morally significant loss on any other person r at w', then a' is at least as good for r at w' as a is for r at w.

Thus, as between two otherwise morally indistinguishable acts, OPPP4 requires agents to impose the lesser and shallower morally significant loss on one person rather than the greater and deeper morally significant loss on someone else. It is left open, as before, whether failing to bring someone into an existence worth having imposes a morally significant loss.

OPPP4 can be viewed as an extension of OPPP1. Whereas the usual Pareto idea applies when agents have the alternative of making things better for a *particular person* without making things worse for anyone else, OPPP4 applies when agents have the option of making things better for p than they are for a *distinct* person q.

OPPP4 governs *very* simple tradeoff scenarios. We actually will not need OPPP4 until we turn to the problem of late abortion in Chapter 3. But we do face a handful of somewhat more difficult tradeoff scenarios in this Chapter 2. And in that connection we will consider what sorts of tradeoff principles seem plausible. But those principles, unlike OPPP4, are just a bit too controversial to be included in the otherwise plausible permissibility theory.

More generally, other principles will be considered as needed, with any controversies noted as we go. The general assumption is that principles that become part of the otherwise plausible permissibility theory will take a *comparative* approach to determining permissibility. Just like OPPP1–OPPP4, they will evaluate the permissibility of a given act by reference to *just how dreadful the alternatives to that act are*. If all the alternatives to a given act are *just as dreadful* as the one act, then the one act is permissible. If the one act is *more dreadful* than some alternative act, then the one act is wrong. But by *more dreadful* here

I do not mean that one alternative act generates less wellbeing *in the aggregate* than the other.[52] Rather, I am just recognizing in an informal way that alternative acts often have certain features that tell us something about the permissibility of any one act.

In addition, we can freely appeal to various principles that we can describe as *standard deontic axioms*. Such axioms plausibly include: if an act is obligatory, it is permissible; if an act is obligatory, none of its alternatives are permissible; and if each alternative act is wrong, then any one act must be permissible.[53] Moreover, if the otherwise plausible permissibility theory happens to be relevantly *complete*, and if no alternative exists for agents that, according to that theory, shows that the one act is *wrong* – if, that is, *each such alternative* is *at least as dreadful as the one* – then we will have no basis for thinking that the one act *is* wrong. And we can then say that that act is permissible.

We can think of such deontic axioms as part of any otherwise plausible permissibility theory. But we need not do so. It doesn't matter. If they are axioms, we can count on them being there for us in any event. And – of course – we must make sure that our otherwise plausible permissibility theory complies with them.

This discussion shows that, while the maximizing account of loss I adopt for purposes here is broad, we will not necessarily end up with an unduly broad concept of *wrongdoing*. Agents often face tradeoffs. They often must choose which of two persons shall incur a loss, and the loss rules we examine in what follows often deem *both* such prospective losses morally significant. But it won't follow that the acts that impose those losses will *both* be wrong. Instead, as long as the tradeoff between the persons in conflict is made in an *appropriate* way – is made, that is, in the way that the *otherwise plausible permissibility theory* directs – what the agent has done will be deemed permissible.

2.2.5 *The Nonidentity Problem.* It seems indisputable that the choices we make today regarding the *wellbeing levels* to be accorded to future persons can have a very significant impact on the *identities* of those future persons. This fact raises the specter of the *nonidentity problem.* Some theorists might consider it a mistake to rely so heavily on an account of loss that requires for loss that agents create less wellbeing for a person when they could have created more wellbeing for that *same identical* person. They might think that the nonidentity problem means that many cases in which we *want* to say (and naively think we *can* say and *need* to be able to say, if we want to make loss so central to an account of wrongdoing) that a loss *bears on* the permissibility of a given act are cases that, on closer inspection, do not,

[52] Thus we do not assume that the dreadfulness of a given alternative is just a matter of the aggregate wellbeing, or amalgamated overall good, we see under that alternative. To assume that would be to assume that the loss of leaving a person out of an existence worth having is itself a *morally significant* loss, and that is not something that the otherwise plausible permissibility theory should make any assumptions about one way or the other. That issue, rather, is what is at stake in this book.

[53] For an account of the fundamentals of deontic logic, see Paul McNamara (2006).

according to the maximizing account I adopt here, involve loss at all. We can bring a *nonidentical* person into a better existence, but we cannot (or so the nonidentity problem asserts) bring that *very same person* into a better existence.

For two reasons I only briefly address the nonidentity problem here. First, I have discussed it in detail elsewhere.[54] And, second, I believe that the nonidentity problem in fact undermines *nothing* we will want or need to say here.

I outline my response to the nonidentity problem in Appendix B below. Very briefly here, I begin by sorting the problems that are collected together under the heading of "nonidentity" into their various problem *types*. Perhaps the most challenging type of nonidentity problem includes Parfit's depletion and risky policy examples, Kavka's slave child and pleasure pill cases and cases involving historical injustice. Even if a better-for-*p* alternative world exists and is accessible to agents, what I call the *can't-expect-better* problem (correctly) notes that the *chances* that agents would have been able to bring about that particular world (or one relevantly like it; one that includes a better off *p*) in place of a world in which *p* never exists at all are miniscule. But isn't it *better for p* to have the flawed existence than to have the miniscule chance of an unflawed existence? And doesn't that mean, under a maximizing account of loss, that *p* has incurred no loss at all? But isn't the choice under scrutiny nonetheless clearly wrong?

A second type of nonidentity problem – the *can't-do-better* problem – applies in the case where there is no world that is both better-for-*p* and accessible to agents. A good example of this second type of problem is the case where a child is born with a serious genetic disorder we do not yet have the ability to correct or to completely palliate, and the child's wellbeing level is unavoidably diminished compared to – say – the norm. Isn't it nonetheless better from that child's own perspective to have the flawed existence rather than none at all? (We assume that the child's life remains worth living – that we are not dealing here with the *genuinely wrongful* life.) And doesn't that mean that the child incurs no loss? But, when agents had the alternative of bringing a nonidentical but better off child into existence in place of the one, isn't the choice to bring the impaired child into existence clearly wrong?

Having divided the cases, I then argue that we can resolve "the nonidentity problem" – that is, that we can resolve *each type* of nonidentity problem on its own terms – so long as we take them *one-by-one* and *not both at the same time*. And I argue that we can do so *without* embracing Inclusion.

Thus, the can't-expect-better version of the nonidentity problem is, I argue, the product of what I elsewhere call the *nonidentity fallacy*.[55] I ought to have said *fallacies*. It is a fallacy to think we can infer that no loss is imposed on a particular person from the fact that the *actual* wellbeing level generated for a person by one act is greater than the *expected* wellbeing level generated for that same person by

[54] See Roberts (2009a, c, 2007).

[55] See Roberts (2009c, 2007).

the other act. When we rewrite the argument, so that we are just comparing one *expected* wellbeing level against another *expected* wellbeing level, we happen onto still another fallacy: the fallacy of thinking that just because a given wrong act has been performed, and the future has unfolded in a way that includes a particular child, the probability, at that critical time just before choice, is more than miniscule that that child will exist, given that choice. The fact is that what we have is a wash: the odds against any particular child coming into existence are *very great*, whether the wrong act is performed or a better act is performed in its place.

The can't-do-better problem involves the rare case where the choice to bring an unavoidably impaired child into existence maximizes wellbeing for each and every person who does or will exist. We may agree, depending on which loss rule we at the end of the day adopt, that such cases involve no morally significant loss at all – none incurred by those who do or will exist, and none incurred by any merely possible person. But it seems to me that we should also think, about those same cases, that it is not clear that a wrong has been done.

More typically, of course, the choice to bring an unavoidably impaired child into existence involves losses that will count as morally significant whatever loss rule we at the end of the day adopt. We can, that is, easily find what we can think of not as a *procreative* loss but rather as a *distributive* loss – though still a bona fide *loss*. That loss can just as easily be incurred by the genetically impaired child or by that child's already-existing or later siblings or by still others. Depending on the facts, that loss can then easily support the result that what the agents have done is wrong.

2.3 The Basic Case

2.3.1 The *Basic Case* underlines what we like about Exclusion (Case 2.1). It parallels the case of the "terrible spat" between the existing dog (gingham or not, as long as it qualifies as a *person*) and the merely possible cat.

Suppose we must choose between acts a1 and a2, where a1 can be counted on[56] to bring about:

[56] In general, we have little control over what possible future, or world, in all its specificity, our choice will produce. For one thing, there may be no settled matter of fact that there is any one particular way the future will unfold if we happen to choose a1 in place of a2. For another, there may be many different ways of realizing a1 itself, and those different ways might themselves trigger different futures. Still, in some cases, we can control whether one particular outcome or another is *part of* whatever world it is at which we perform one particular act rather than another. Let's suppose that we are in that kind of case here.

> A person Ann's having an existence that is well worth having, and a possible person Bob's never coming into existence at all and for that reason accruing an overall wellbeing level of exactly zero.[57]

And a2 can be counted on to bring about:

> Ann's coming into existence and having an overall wellbeing level of exactly zero, and Bob's coming into an existence that is just as worth having as Ann's is under al.

We can diagram the case as follows. Where (i) natural numbers represent overall wellbeing levels[58], (ii) a1 and a2, which exhaust our alternatives, are performed at (arbitrary) worlds w1 and w2 where they respectively bring about the outcomes just described, (iii) boldface means that a specified person *does or will exist* at the specified world and italics, followed by the asterisk, that the specified person *never exists at all* at the specified world; and (iv) no one other than Ann and Bob have anything at stake in whether a1 or a2 is performed:

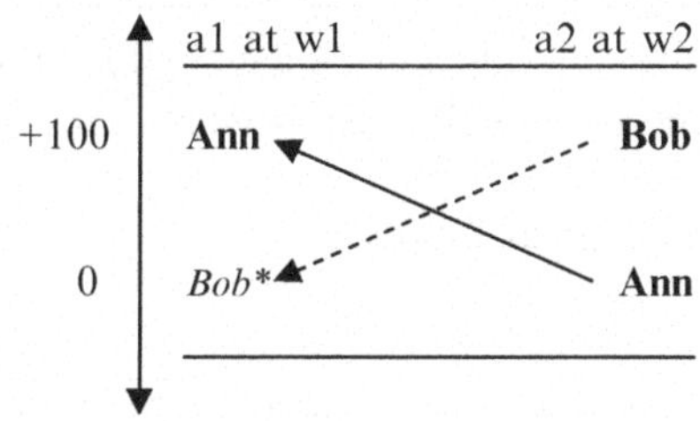

Case 2.1 Basic Case

a2 makes things much worse for Ann than a1 does. She, accordingly, incurs a loss under a2 but not a1. There is, moreover, at least a case to be made – a case that, as we have seen, is taken for granted for purposes here – that a1 makes things much worse for Bob than a2 does, and that Bob, accordingly, incurs a loss under a1 but

[57] That Bob's overall wellbeing level is zero at any world where Bob never exists at all is an assumption that I make for purposes here. More generally, I assume that that one's overall wellbeing level at any world where one never exists at all is zero. See parts 2.2.1–2.2.3. This assumption is, though controversial, one we should be willing to make for purposes here. For something like it plausibly will be part of the case for the claim that a2 makes things better for Bob than a1 does and that a1 accordingly imposes a *loss* on Bob. That case is at least cogent enough that we would not want to rest our own account of what ought to be done on the assumption that it can be safely rejected. Nor do we want to beg the question against Inclusion.

[58] These natural numbers, with the exception of zero, need have their ordinal value only. By *overall wellbeing levels*, I mean *lifetime* wellbeing levels, taking into account all the benefits and all the burdens that ever come to the specified person at the specified world. More generally, *wellbeing* for purposes here means whatever it is that makes life so precious to the one who lives – not just subjectively precious, but truly of value, to that person. It seems to me that a capability approach to value will be important in defining wellbeing. See for example Sen (1992). But we do not want to overlook that the subjective component is important, too – that capability combined with joy is better for a person than capability alone (even if we can't figure out a way to say that joy itself is a part of capability). For purposes here, however, I set aside whether this way of looking at wellbeing is plausible. For a more recent discussion, see Feldman (2009).

not a2. After all, a2 creates more wellbeing for Bob than a1 does, since Bob by hypothesis has an existence "well worth having" at w2 and has nothing at all – no benefits; no burdens – at w1.[59]

Our intuitions about what ought to be done in the Basic Case seem clear: in some way *Ann must come first*. a1 is obligatory, and a2 is wrong. We ought to sacrifice the interests – such as they are – of the *possible* person Bob since doing so is *necessary* to protect the interests of the *existing* Ann. This is not to claim that we will not, by dint of argument at the end of the day, find ourselves forced to abandon intuition and adopt another view. This is just to claim that, starting out, we intuitively think that a1 is obligatory and a2 is wrong.

2.4 Exclusion Alpha

2.4.1 The *truly actualist* form of Exclusion – what I call *Exclusion Alpha* – provides that only those persons who do or will exist in what is the uniquely *actual* world matter morally.

We can state Exclusion Alpha as follows.

Exclusion Alpha:

Losses incurred at the *actual world* w or at any other accessible world w' by any person p who does not and will not exist *at* w – including the loss incurred by p when agents fail to bring p into existence – have *no moral significance* for purposes of determining the permissibility of any act a performed at w that imposes those losses or any alternative act a' performed at any world w' that avoids those losses.

A loss incurred at the *actual world* w or at any other accessible world w' by any person p who does or will exist at w is morally significant for purposes of determining the permissibility of an act a performed at w that imposes that loss and any alternative act a' performed at any other world w' that avoids that loss.

---

[59] Again, I assume that the wellbeing level a person has a world where that person never exists at all is just zero. See part 2.2.1–2.2.3 above.

Nonidentity worries can arise at this point. The supposition that the act a2 will eventually lead to the conception and coming into existence of any particular person *Bob* (as opposed to any one of many, many nonidentical others) will hold only on the rarest of occasions. More typically, we can be confident only that an act will lead to the existence of *some person or another*. To insure that the discussion that follows applies in both sorts of situations, we can understand the term "Bob" itself to function as a constant, not a name. Thus, we are saying that a2 will produce a certain sort of outcome at the world w2 where a2 is performed. That outcome will include some person or another – call that person "Bob" – who will have an existence well worth having; and that same person will not exist at all if a1 is performed instead. At the end of the argument, we can then discharge the constant "Bob" in accordance with the normal rules for existential derivation without any adverse impact on the conclusion we mean to draw – whether that conclusion is that a1 is obligatory in view of its effect on Ann, or a1 and a2 are both permissible in view of their effects on Ann and some other person, or something else entirely.

According to Exclusion Alpha, the effects of our acts on actual persons and actual persons alone are to be taken into account in determining what we do. Their losses, and their losses alone, have moral significance for purposes of evaluating the acts that impose those losses and the alternative acts that avoid those losses.

2.4.2 Exclusion Alpha won't tell us anything about the Basic Case until we add one additional fact. We need to say which world is *actual.*

Let's suppose that w1 is the uniquely *actual* world – and that a1, accordingly, is the act *actually* performed and that Ann *actually* exists while Bob remains *merely possible* relative to the actual world.

Let's then begin by looking at the loss Bob incurs at w1. Since w1 is actual and Bob never exists at w1, Exclusion Alpha implies that the loss Bob incurs at w1 has *no* moral significance for purposes of evaluating a1. Nor does that loss have any moral significance for purposes of evaluating a2 *even though Bob exists at* w2. Bob's loss at w2 is thus, according to Exclusion Alpha, devoid of moral significance across the board.

In contrast, since Ann exists at the actual w1, Exclusion Alpha implies that the loss she incurs at w2 has *full moral significance.* It *bears on* the permissibility of *both* a1 and a2. It *counts against* a2. In contrast, a1 altogether avoids that loss on Ann's behalf – not just in a vacuous way, by leaving Ann out of existence altogether, but rather in a way that makes things *better for Ann.* The otherwise plausible permissibility theory – specifically, OPPP1 – thus directs that a2 is wrong. Moreover, since a1 imposes no morally significant loss on anyone at all – on any existing, future or merely possible person; no loss at all on Ann, and no morally significant loss on Bob – OPPP2 instructs that a1 is permissible.[60] Standard deontic axioms tell us more: a1 is not just permissible, but *obligatory*, given that a2 itself is wrong.

And that result is a plus for Exclusion Alpha. At the very least, Exclusion Alpha comports very nicely with our original intuitions about the Basic Case.

2.4.3 But a slight change in the hypothetical turns this moral evaluation on its head. Before we added the fact that w1 is actual. Let's now instead add the fact that w2 is actual. It is a2 that is actually performed, and it is Ann *and Bob* who actually exist, not just Ann. Exclusion Alpha now instructs that the losses incurred by both Ann and Bob – Bob at w1 and Ann at w2 – bear on the permissibility of both a1 and a2. The case becomes, under Exclusion Alpha, a *wash* – a case involving acts that generate losses that are equally great and equally deep *and* have full moral significance, a case where it just does not matter what the agents choose to do. OPPP3 accordingly implies that a2 is *permissible.* And so is a1.

2.4.4 Exclusion Alpha thus conforms to the clear intuitions we have about the Basic Case only about half-way. When, by hypothesis, w1 is actual and a1 is actually performed, we obtain the intuitive result that a1 is permissible, in fact, obligatory, and a2 is wrong. But when we change the hypothetical and instead suppose that w2 is actual and a2 is actually performed, we suddenly obtain the result that a1 and a2 are *both permissible.*

But *does* the new hypothetical really trigger a new intuition? Does a2 – and it's the same old act, just situated within a new case; it's the same old act, now made

[60] We could, but need not, think of such deontic axioms as part of the otherwise plausible permissibility theory. See part 2.2.4 above.

actual when before we stipulated that it was left *unperformed* at the actual world – now seem *permissible*? We might just adore Bob, if he exists and we, too, exist. But our intuition as to permissibility is surely exactly what it was in the original case: a2 is wrong.

Exclusion Alpha has already committed itself to the view that if w1 is actual and a1 is actually performed – that is, if a2 is left *unperformed* – then a1 is obligatory and a2 is wrong. It is hard to understand how the mere *performance* of a2, which we just said was wrong if *unperformed*, could make a2 *right*.

The more plausible view is surely that the permissibility of an act does not depend on whether that act is actually performed. Put another way, Exclusion Alpha violates what Wlodek Rabinowicz calls the *Principle of Normative Invariance*.[61]

2.4.5 What we are seeing in Exclusion Alpha can be described as a *conceptual instability*, a situation in which the evaluation undergoes a seismic shift as a result of something that – intuitively – has more to do with history or metaphysics (was the dastardly act in fact performed or not?) than morality.

Now, the supposition that w1 is actual and that a1 is actually performed, and the alternate supposition that w2 is actual and that a2 is actually performed, cannot both be correct in any one case. There is only one *actual* world, and the performance of a1 precludes the performance of a2. Thus we are really talking about two distinct cases – the Basic Case, where w1 is actual, and an alternate version of that case, where w2 is actual instead. So the conceptual instability that we see does not itself suggest any logical inconsistency.

But that does not mean that the problem is not a very serious one. We have already said that Exclusion Alpha conforms only about half-way to the intuitions we have regarding the Basic Case. And now we see that Exclusion Alpha reads tea

---

[61] According to that principle, the permissibility of an act does not depend on whether that act is performed. The principle is attributed to Rabinowicz by Eric Carlson (1995), p. 100. The difficulty with theories that fail to conform to the Principle of Normative Invariance is, according to Carlson, that they are not "action-guiding." Carlson (1995), p. 101. It is not beyond conception that a plausible permissibility theory may sometimes for any number of reasons fail to guide action. A theory can, for example, be plausible without being *complete*. Moreover, actual value forms of consequentialism may be less adept at guiding action than are their expected value analogs. But it is probably a mistake to reject actual value theories across the board just on the ground that they often fail to guide action; after all, expected value approaches often themselves fail to point us precisely in the direction of what we ought to do or permissibly may do. In my own view, then, the Principle of Normative Invariance goes still deeper: it underlines that whether an act is performed or not is not itself a morally salient property. A *wrong* but unperformed act seemingly *can't* be made *permissible* by nothing more than the agent's performance of that act. Many facts are not pertinent to moral permissibility; the work of constructing a moral theory will be successful or not depending on our ability to attend to the facts that are pertinent and, just as important, *to disregard the rest*. Before we can accept, in other words, that performance is pertinent to permissibility, we need an argument – or at least a handful of cases that compel us to appreciate the connection on an intuitive basis. Certainly, however, none of the cases we shall be examining here make that connection. The cases under consideration here, if anything, suggest just the reverse: that what is going on in these cases has nothing to do with whether the act is actually performed or not. For critical discussion of the Principle of Normative Invariance, see Frances Howard-Snyder (2008), pp. 1–15.

leaves that we ourselves cannot connect with what we are trying to explain. It thinks performance is important, and we can't see how that is so.

2.4.6 There's still another problem with Exclusion Alpha. It leads us to the result that either act, if *actually* performed, is permissible. But the whole point of Exclusion Alpha was to *put Ann first* – to assert, that is, that it *does* matter in this kind of case whether we avoid the loss on Ann's behalf or not, that it's not "just as good" to avoid the loss on behalf of Bob. This means that Exclusion Alpha isn't doing the work that it was meant to do. That in itself does not mean that Exclusion Alpha is *false*. Perhaps it was our initial intuitions about the Basic Case that were defective. But before we decide it's the intuitions that are defective and not Exclusion Alpha, we should see whether there is another way of making sense of those intuitions.

2.5 Double Wrongful Life

2.5.1 Other cases challenge Exclusion Alpha as well.[62] In *Double Wrongful Life*, the agent's choices are limited to bringing one or the other of two *wrongful lives* – lives, that is, that are *less* than worth living – into existence (Case 2.2).[63]

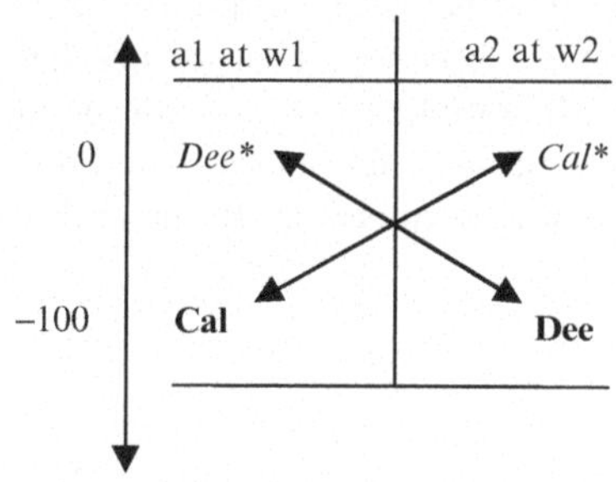

Case 2.2 Double Wrongful Life

Let's first suppose that w1 is the uniquely actual world and that a1 is actually performed. Cal thus actually exists and is accorded a wrongful life – a wellbeing level of –100. According to Exclusion Alpha, since Cal alone exists at w1, the loss Cal incurs at w1 alone bears on the permissibility of a1 and a2. The loss Dee incurs at w2 is to be screened out of the analysis altogether. a2 thus avoids a morally significant loss on behalf of Cal and makes things better for Cal – all without imposing a morally significant loss on anyone. On those facts, OPPP1 implies that a1 is wrong. OPPP2, moreover, instructs that a2 is permissible. Standard deontic axioms tell us still more: that a2 is obligatory.

[62] The objections against Exclusion, in both its Alpha and Beta forms, that I explore here have been discussed elsewhere. See Arrhenius (2009) and Caspar Hare (2007). See also Parsons (2002), pp. 135–47 and, generally, Parfit (1987), part four.

[63] Case 2 is based on Hare's Jack and Jane case, which he describes as "negatively symmetrical." Hare (2007), p. 504.

2.5.2 But our intuitions surely go in the other direction. The very symmetry of the case tells us that a1 and a2 should be similarly evaluated. Either both acts are wrong or they are both right.[64] But it seems that moral law must always give agents at least one permissible alternative. So intuition alone leads us to think that a1 and a2 are *both* permissible.

2.5.3 There's an even deeper problem. Let's now change hypotheticals. Let's now suppose that it is a2 that is actually performed rather than a1. Exclusion Alpha now, by a parallel analysis, generates the result that *a2* is wrong and *a1* obligatory.

But we have already said that, in the original version of the case, the version in which w1 is actual, that a1 is wrong and a2 obligatory. And that means that, whether the agent actually performs a1 or actually performs a2, what the agent does is wrong *and* what the agent could have done instead is permissible.

2.5.4 If this looks like an inconsistency in Exclusion Alpha, it isn't. For we are really dealing with two distinct cases, one that includes the supposition that w1 is actual, and one that includes the supposition that w2 is actual.

However, we do face here exactly the same sort of conceptual instability we saw in the Basic Case. If a1 is actually performed, a2 looks permissible. But the second we shift cases, and suppose instead that it is a2 that is actually performed rather than a1, the evaluation of a2 also shifts.

2.5.5 One last point. It's true that both variations on Double Wrongful Life leave the agent with a permissible alternative. That means that Exclusion Alpha avoids the classic moral dilemma. It complies, that is, with the idea that moral law always gives agents at least one permissible alternative. The problem is they can't *perform* that alternative and have it *stay* permissible. In the end, whatever they actually do is sure to be wrong.

2.6 Addition Plus

2.6.1 Still another problem case for Exclusion Alpha is *Addition Plus*.[65] Addition Plus is like the Mere Addition Paradox in that an additional person whose life is worth living is added to the mix.[66] The "Plus" is that adding the additional person here makes *someone else better off.*

---

[64] The differences between a1 and a2 constitute *merely reversing changes*. See generally Peter Vallentyne (2000), pp. 1–19.

[65] I have elsewhere argued that this sort of case shows that such overly simplistic sufficient conditions on impermissibility – however person-based in nature they may be considered to be – must be rejected. See Roberts (1998), pp. 63–64.

[66] Addition Plus raises many of the same issues as one of the problem cases that Parfit and Singer discussed in connection with a proposal explored by Singer. That proposal was an attempt to articulate an alternative to the Totalist treatment of additional person cases generally. See Singer (1976), p. 97, and Parfit (1976), p. 105. See also Parfit (1987), pp. 419–41. More recently, Singer articulated and rejected what he called the *Prior Existence View*. See Singer (1999), pp. 102 ff. For discussion of the Prior Existence View, see part 2.11 below.

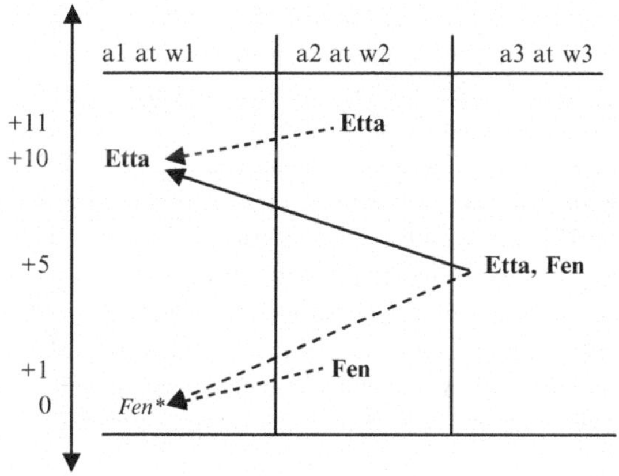

Case 2.3 Addition Plus

Let's begin with the supposition that w1 is the actual world and a1 is actually performed. Then, according to Exclusion Alpha, since Etta alone exists at w1 and w1 is actual, Etta's losses alone have moral significance for purposes of evaluating each of the agent's alternatives – not just a1 but a2 and a3 as well. OPPP1 implies, accordingly, that a1 and a3 are both wrong; and OPPP2 that a2 is permissible. Standard deontic axioms tell us, as well, that a2 is obligatory. After all, a2 avoids not one but two morally significant losses on Etta's behalf, and makes things better for her than either a1 or a3 does. It's true that Fen incurs a loss at w2. That he does is established not by comparing how well Fen fares in w2 against how well he fares in w1, but rather by comparing how well Fen fares in w2 against how well he fares in w3. The former comparison demonstrates no loss at all; the latter, a very great and very deep loss. But since Fen never exists at w1, Exclusion Alpha implies that that very great and very deep loss is *devoid of moral significance*.[67]

2.6.2 But the results that a1 is wrong and a2 obligatory intuitively seem false. We cannot really think that it is wrong for agents to refrain from bringing a new person into an avoidably bad existence all in order to secure some slight gain for one person. Since it is part of the case that agents have the option, at w3, of doing so much more for Fen, we cannot really think that it is obligatory, or even permissible, for agents to bring Fen into existence *and do so much less for him*, all to secure some slight benefit for Etta. That it would have created additional wellbeing for my

[67] Now, we may well consider wellbeing more appropriately distributed between Etta and Fen under a3 than under a2. But the tradeoff principles an otherwise plausible permissibility theory can be expected to include will never require the agent to impose a morally *significant* loss on one person – Etta in w2 – so that the agent can avoid imposing a morally *insignificant* loss on someone else – Fen in w1.

child to have her own private live-organ donor, or slave, does not make it wrong for me *not* to have brought an additional person into existence *and to have avoidably treated that person badly*.[68] There are limits to what I may do even when the person I am doing more for is *my own child*.

2.6.3 Of course, some philosophers – inspired, perhaps, by a total, or pluralistic, form of aggregative consequentialism, and notwithstanding the point that both Totalism and Pluralism are quite at odds with Exclusion Alpha[69] – may want to press the argument that a1 *is* wrong. I don't think a1 is wrong. But if it is, we should note that *even the Totalist* agrees that any plausible view must *at least* take Fen's loss at w2 into account as it makes its way toward a determination of a1's permissibility. Fen's loss at w2 at least *bears on* the permissibility of a1. That modest point is, however, one that Exclusion Alpha must reject in the case where w1 is actual.

2.6.4 Many of us have the following intuitions. (A) Etta is not compelled to suffer just so that the merely possible Fen can come into existence and be treated well, as Fen is under a3. But (B) *if* Fen is brought into existence at all, the fact that a3 is a perfectly viable alternative way of bringing Fen into existence means that a2 itself is far too flawed to establish that a1 is wrong. Affirming both (A) and (B) means that a1 is (at least) permissible. But it also means that the loss Fen incurs at w2 *bears on* our evaluation of the permissibility of a1 – and that that is so, notwithstanding the fact that Fen never actually exists at w1. By counting *against* a2, Fen's loss at w2 in a roundabout way thus counts *in favor of* a1.

Fen's loss at w2 thus *seems*, at least, enough to make the otherwise obligatory but nonactual a2 wrong and the otherwise wrong but actual a1 permissible. But if that is correct, then we shall have no choice but to reject Exclusion Alpha, which insists that Fen's loss at w2 has *no bearing at all* on the permissibility of a1.

2.6.5 Nils Holtug suggests that views like Exclusion Alpha, by virtue of the claim that persons who are merely possible relative to the actual world *never* matter morally, have more in common with "racism, sexism and speciesism" than we would like.[70] Holtug's language seems a bit over the top until we come to understand

---

[68] The nonidentity problem arises here. It may seem to undermine the assertion that a3 is better for Fen than a2 is. Can we really make it *part of* Addition Plus that agents have the genuine alternative in a3 of creating additional wellbeing *for Fen*, the very same, identical person who is made to suffer under a2? The simple answer here is that it's part of the hypothetical that agents in fact do have that very option. The more interesting answer is that the very type of nonidentity problem that might be used to challenge the hypothetical is in fact based on a fallacy. See part 2.2.5 above and Appendix B.

[69] See note 36 above (description of Pluralism).

[70] Holtug (1999), p. 26. It is not necessarily that we have Kantian objections against, for example, slavery. It might just as well be that we are consequentialists of one sort or another and think that the option of bringing someone into such an impoverished existence – with so little justification; just 1 unit of wellbeing for the already very well-off Etta – when we had the option of *doing so much more* for that same person cannot establish that the alternative of leaving that person out of existence altogether is wrong.

just what a fully-charged commitment to existing and future persons obligates us to do to the merely possible.

The difficulty is not at all that Exclusion Alpha on occasion obligates us to *leave the merely possible out of existence altogether* – when, for example, doing so makes things better for some existing or future person without making things worse for anyone at all. The difficulty is that Exclusion Alpha on occasion also makes it our obligation to bring the merely possible into existence and *unjustifiably* create less wellbeing for those same ("identical") persons when we could have created more. But surely that is not an obligation we in fact have.[71]

2.6.6 Caspar Hare describes still another case that shows that Exclusion Alpha struggles with nonactual acts whose adverse effects are limited to nonactual persons. Exclusion Alpha – what Hare calls *Strong Moral Actualism* – seems to instruct that the "genocidal adventures of nonactual dictators" are permissible, or even obligatory, in cases where the victims of those adventures are themselves nonactual.[72] Exclusion Alpha leaves the otherwise plausible permissibility theory with no grounds for declaring such "genocidal adventures" wrong. But we think that they are.

2.6.7 That we intuitively consider wrong the terrible things Darth Vader did to *his own son* might seem reason enough to reject Exclusion Alpha. In defense of Exclusion Alpha, however, it might be argued that we should not be too concerned with how the evaluation of *nonactual* acts comes out. It may seem that we should not be particularly troubled when the theory fails to produce satisfying results for acts *whose victims are merely hypothetical.* It may seem that what is important is that the theory produce satisfying results for acts *whose victims are actual.* And, indeed, had the "genocidal adventures of nonactual dictators" in fact taken place, Exclusion Alpha, in combination with any otherwise plausible permissibility theory, would no doubt have judged those adventures wrong since their victims would then have been actual.[73]

This defense of Exclusion Alpha can be put another way. Exclusion Alpha comes with a caveat. If we *had* actually done as we ought – if we had actually brought those merely possible persons into existence *and* treated them badly – then those persons would not have been *merely possible* but rather *actual.* If, in other words, we revise the case in such a way that the act *actually* performed, and the class of persons who do or will exist at the uniquely *actual* world shifts – if, in other words, we *had* done as we ought – then Exclusion Alpha would have declared what we have done wrong.

The caveat thus insures that the "genocidal adventures of nonactual dictators" *would have been wrong* had those dictators and their bad acts been actual rather than not.

Similarly, it's true that, in Addition Plus, where a1 is actually performed and Etta and only Etta accordingly matters, a2 is obligatory. But the caveat insures that, had a2 actually been performed in lieu of a1, a2 *would have been wrong.*

---

[71] Holtug (1999), p. 26.

[72] Hare (2007), 503. But see Parsons (2002), pp. 135–47.

[73] See generally Parsons (2002), p. 145.

2.6.8 But it is not clear whether this caveat rescues Exclusion Alpha from its own excesses or reveals just how deeply flawed Exclusion Alpha really is. The difficulty is that Exclusion Alpha *cannot be contained to questionable evaluations of nonactual acts and the nonactual losses they impose*. The difficulty extends to how *actual* acts, imposing *actual* losses, are evaluated as well.

Addition Plus makes this point. Part of the reason that the *actual* a1 intuitively seems permissible in Addition Plus, notwithstanding the *actual* loss it imposes on Etta, is what the *nonactual* a2 does to the nonactual Fen. When Exclusion Alpha instructs us to exclude Fen's interests from our calculation of the permissibility of a1, it forces us to say that the additional wellbeing a2 creates for Etta is *sufficient* to show that a1 is wrong. Thus a1, according to Exclusion Alpha, is not simply wrong *if* performed. It's both performed *and* wrong. The upshot is that the deficiency we see in Exclusion Alpha cannot be contained to nonactual acts and the nonactual losses they impose.

Thus, the caveat shows that, had another world and another act been actual, we would have been able to obtain, under Exclusion Alpha, a more plausible permissibility result. That's all well and good. The problem is that it doesn't in the least change the fact that we have already drawn from Exclusion Alpha a permissibility result that is not plausible at all: that the *actual* a1 is wrong.

2.6.9 The conceptual instabilities that we earlier noted in connection with Exclusion Alpha are repeated in Addition Plus. Suppose that w1 is the actual world and that it is a1 that is actually performed. As a simple conceptual matter, it is hard to understand how our performing the nonactual act a2 that is itself, if unperformed, permissible or even obligatory – *as we ought if it's obligatory* – can make that act *wrong*. The better view seems to be that, if mere performance makes the act wrong, surely it was never *genuinely obligatory* to begin with. But if we take that view, we are forced to reject Exclusion Alpha.

2.6.10 There is a related problem. Addition Plus, like Double Wrongful Life, underlines that Exclusion Alpha cannot be counted on to give us a *way out of wrongdoing*. *Not* bringing the merely possible Fen into existence and treating him badly is wrong in virtue of its being bad for Etta. But *bringing* Fen into existence and treating him badly – had we actually done that instead – would have been just as wrong. We again have a case in which the agent can *do no right*. This is not technically a moral dilemma, but it doesn't really adhere to the spirit of "ought implies can," either. Whatever the agent does, he or she has a morally permissible alternative. The problem is that the agent can't *perform* that alternative and have it *stay* permissible.

2.6.11 Why is Exclusion Alpha failing in these cases? The obvious point is that Exclusion Alpha is *too exclusive*: it is screening out morally significant losses. It thus fails to recognize in Double Wrongful Life that the loss Dee incurs at w2 is what makes a1 *permissible*; her loss counts, not just *against* a2 in the case where a2 is actually performed, but, in a roundabout way, *in favor of* a1. Similarly, even if Cal never actually exists, the loss he incurs at w1 is what makes a2 *permissible*. Addition Plus makes the same point. The loss Fen incurs at w2 clearly at least *bears on* our evaluation of both a1 and a2 even if a1 is the act actually performed and Fen remains merely possible. It is that loss that makes a2 clearly wrong – and, in a roundabout way, a1 plausibly *right*.

2.6.12 But the more subtle point is that Exclusion Alpha is *not exclusive enough*. At certain points, it allows the merely possible to exert more of a moral pull on what the agent does than they ought. According to Exclusion Alpha, all that has to happen, for the merely possible to matter morally, is for the supposition regarding which act is actually performed to be modified. At first glance, that feature seems desirable – or at least innocuous. If the *merely possible* are *made actual* by changing the facts of the case, then surely they will then come to matter morally. If nothing else, surely we can agree that *all actual people always matter morally*.

I believe, however, that this is one of those many cases in which a superficially obvious truth is false. It is a mistake, that is, to think that actual persons matter morally in some *necessary* way – to think, that is, that *all* their losses have moral significance. The Basic Case makes that point. Where a1 is actually performed, Exclusion Alpha functions just as it should. It goes awry when we shift to the supposition that a2 is actually performed. Then Exclusion Alpha abruptly assigns moral significance to *all* of the losses incurred by Bob *wherever* he incurs them, *including* at w1 where he never exists at all. But the fact that Bob is actual hardly suffices to make a2 permissible. We can *just adore* the actual Bob but still consider what has been done to Ann wrong and still insist that what we ought to have done was leave Bob out of existence altogether.

Perhaps we are pleased that we ourselves happened to have been conceived against all odds and then born.[74] I certainly am. But that is not enough to show that we *matter morally in some necessary way* – that all of our losses, wherever we happen to incur them, have full moral significance or indeed any moral significance at all. That is not enough to show that the loss we would have incurred had we never existed at all *counts against* whatever it was our parents might have done in lieu of bringing *us* into existence.

The same point holds for the Basic Case. Bob will exist if a2 is performed. But whether he exists or not, the loss he incurs at w1, where he never exists, has no moral significance. The puzzle, of course, is how to say that *without* implying that it is permissible or even obligatory, in Addition Plus, to bring Fen into existence and then treat him badly.

2.6.13 The cases thus suggest that Exclusion Alpha is both too exclusive *and* not exclusive enough. I think that means that it's not going to work either to loosen *or* tighten Exclusion Alpha. It means that we need a mechanism for doing both at the same time. But is that a mechanism either Exclusion Beta or Inclusion can provide?

[74] "Imagine a speck of dust next to a planet a billion times the size of the earth. The speak of dust represents the odds in favor of your being born, the huge planet would be the odds against it. So stop sweating the small stuff. Don't be like the ingrate who got a castle as a present and worried about the mildew in the bathroom. Stop looking the gift horse in the mouth – remember that you are a Black Swan." Taleb (2007), p. 298.

2.7 Exclusion Beta

2.7.1 For Exclusion Beta, the critical fact is not who *is* actual and how an act and its alternatives affect those persons, but rather who *would be* "actual" if the act we aim to evaluate had been performed. Under Exclusion Beta, just how a *nonactual* act affects *nonactual* persons becomes just as important as how an *actual* act affects *actual* persons. The only restriction is that those nonactual persons must exist, at least eventually, *under that nonactual act* for their losses to have moral significance for purposes of evaluating that act.

Exclusion Beta can be put as follows.

Exclusion Beta:

Losses incurred at any world *w* or at any other accessible world *w′* by any person *p* who does not and will not exist at w – including the loss incurred by *p* when agents fail at *w* to bring *p* into an existence worth having – have *no moral significance* for purposes of determining the permissibility of any act *a* performed at w.

Losses incurred at any world *w* or at any other accessible world *w′* by a person *p* who does or will exist at *w* are morally significant for purposes of determining the permissibility of an act *a* performed at w.

A person thus matters morally for purposes of evaluating a particular act *if and only if* that person does or will exist at the world at which that act is performed. This means a person who counts as merely possible relative to the *actual* world may still matter morally. That person will matter morally if he or she does or will exist at the world at which the act under scrutiny is performed.

2.7.2 Exclusion Beta starts out well enough in the Basic Case. At least: Exclusion Beta evaluates a1 just as Exclusion Alpha does when combined with the supposition that w1 is the actual world. (And Exclusion Beta doesn't need the supposition.) Since a1 is the act under scrutiny and Ann and Ann alone exists if a1 is performed, Exclusion Beta implies that the loss Ann incurs at w2 is morally significant for purposes of evaluating a1 but the loss Bob incurs at w1 is not. The analysis under the otherwise plausible permissibility theory proceeds just as it did under Exclusion Alpha. We obtain the result that a1 is obligatory – a result that perfectly comports with our intuitions about the Basic Case.

2.7.3 But when we turn to a2, Exclusion Beta quickly runs into trouble. Both Ann and Bob exist if a2 is performed. Exclusion Beta thus implies that the losses incurred by Bob at w1 as well as by Ann at w2 have full moral significance.

The analysis now proceeds just as it did for Exclusion Alpha when combined with the supposition that it is a2 that is actually performed. The upshot is that Exclusion Beta, just like Exclusion Alpha, considers the case a *wash*: a case involving acts that generate losses that are equally great and equally deep *and* have full moral significance. The otherwise plausible permissibility theory will direct that a2 is permissible.

That result seems clearly counterintuitive. Our intuition is that Ann must come first. Exclusion Beta now instructs that Ann and Bob both matter morally – more precisely, that all the losses each incurs, whether at w1 or at w2, have full moral significance now that we ourselves have shifted our own focus from a1 to a2.

2.7.4 And there is a second problem. It is not just that it is *counterintuitive* to say that a2 is permissible. It is also that we said just before that a1 is *obligatory*. These are inconsistent results. At least, they flout deontic axioms we may be loathe to reject. To say that an act is *obligatory* is to say that agents must perform *that act and no other*. But to say an alternative act is permissible is to say that agents *may perform that alternative act*. Accordingly, if a1 is *obligatory*, a2 *cannot* be *permissible*.

We said before that Exclusion Alpha was conceptually unstable in a certain way. We evaluated a1 one way given the supposition that w1 was actual and another way given the supposition that w2 was actual. More precisely, we evaluated a1 as obligatory in the version of the case that included the supposition that w1 was actual, and as permissible in the *alternate version* of that case that included the supposition that w2 was actual. But the situation we see in Exclusion Beta is, if anything, a bit worse. We now see a shift in evaluation taking place in the *very same case*. So here we face not "just" a conceptual instability but – taking deontic axioms into account – an out-and-out inconsistency.

The roots of the inconsistency are perfectly clear. Exclusion Beta claims that the class of persons who matter morally can shift, depending on which act happens to be under scrutiny. But we also think that how we evaluate one act constrains how we evaluate another. Deontic axioms – not to mention intuition – forbid our saying in one and the same case both that one act is obligatory and that an alternative act is permissible. Thus, in the Basic Case, Exclusion Beta in combination with the otherwise plausible permissibility theory judges a1 *obligatory* relative to the collection of persons who happen to exist under that act – that is, Ann. But Exclusion Beta *in the very same case* judges an alternative act *permissible* relative to the *distinct* collection of persons who happen to exist under that *distinct* act – that is, Ann *and Bob*. We thus find ourselves saying that a1 is *obligatory* and a2 *permissible*.

2.7.5 Exclusion Beta also stumbles badly in Double Wrongful Life. Thus, when we evaluate a1, we obtain from Exclusion Beta that the loss Cal incurs at w1 and that loss alone has moral significance for purposes of evaluating a1. It's true that Dee incurs a loss at w2. But, according to Exclusion Beta, Dee's loss has no moral significance since it is a1 we are evaluating and Dee never exists at w1. The otherwise plausible permissibility theory – specifically, OPPP1 – accordingly implies that a1 is wrong.

But that result is counterintuitive. By whatever measure, a2 is obviously just as dreadful as a1 is. There is nothing in the very unfortunate a2 that even begins to make us think that the very unfortunate a1 is wrong. We appreciate that a1 has a very bad effect. But it is rash to jump from that fact to the claim that a1 is wrong. We only make the mistake of doing that if we've failed to inspect a1's alternatives in any careful way. When we do, we decide that a1 is permissible.

2.7.6 Things get worse when we turn to evaluate a2. An exactly symmetrical analysis implies that a2 is wrong. But we've just said that a1 is wrong. Of course these results track the results we obtained under Exclusion Alpha. But we've now obtained the results that a1 and a2 are both wrong in the *very same case*, not two versions of a case that differ in the suppositions they make regarding which is the actual world. Exclusion Alpha meant that the agent always had a permissible option, even if that option could not be both actually performed and remain permissible. Exclusion Beta leaves the agent with a classic moral dilemma – a case in which the agent has *no* permissible alternative.

But can that ever happen? A standard deontic axiom says it can't. Now, one might want to challenge that axiom. One would do that by presenting an extreme case in which we are strongly and irremediably pulled to say that both alternatives are wrong. But Double Wrongful Life is just not that kind of case. It's an odd case, but we are not pulled to say that *both* options are wrong. We may think both options are *horrible*, but that is different. That is not enough to make us think that agents have been left with *no permissible alternative* at all.

The discussion would go very differently, of course, if we created a completely new case by giving agents still a *third option*, a3, that would leave both Cal *and* Dee out of existence altogether (or bring them both into existence and give them lives worth living, or etc.). But that's not *this* case – nor would that new case create a moral dilemma. We would just say that a1 and a2 are wrong, and that a3 is permissible and, indeed, obligatory.

2.7.7 Now, Exclusion Beta does have what seem to be some clear advantages over Exclusion Alpha. It seems a plus for Exclusion Beta that it, in combination with the otherwise plausible permissibility theory, can declare nonactual acts whose bad consequences are limited to nonactual persons wrong. Exclusion Beta recognizes that the losses incurred by persons who are *merely possible relative to the actual world* have moral significance for purposes of evaluating any nonactual act under which those persons do or will exist. Exclusion Beta, unlike Exclusion Alpha, can thus declare the "genocidal adventures" of Caspar Hare's "nonactual dictators" wrong.

2.7.8 This feature may make it look like Exclusion Beta does a better job than Exclusion Alpha with nonactual acts and their nonactual victims. Of course, the bare fact that we've already seen very deep conceptual difficulties in Exclusion Beta should make us skeptical that that is so.

What we should like and not like about Exclusion Beta becomes very clear in connection with Addition Plus. Even if w1 is the actual world and Fen never actually exists, Exclusion Beta recognizes that the loss Fen incurs at w2 – the loss Fen incurs not compared to how well he does in w1 (which is not well) but rather compared to how well he does in w3 (which is well) – is morally significant for purposes of evaluating a2. Etta also exists at w2. So the losses she incurs, at both w1 and w3, bear on the permissibility of a2 as well. Clearly, however, the loss Fen incurs at w2 is much *deeper* than any loss Etta incurs at w3. And that is so, whether we are talking about the loss she incurs at w3 relative to w1 or the loss she incurs at w3 relative to w2. The *volumes* of the various losses are pretty close. But, since Fen is left far worse off at w2 than Etta is at w3, the *depth* is very different. On these facts, the otherwise plausible permissibility theory may well deem a2 wrong, given that agents had the option of performing a3 in its place. The thought would be that it does not make the necessary tradeoff between Etta and Fen in an appropriate way.

Now, one might well quarrel with the underlying assumption that any permissibility theory that is both otherwise plausible and more complete will include the idea that agents ought first see to the needs of the least well off, then the next least well off, and so on – will include, that is, a reference to Leximin. I like some of the variations on Leximin, but that's not a debate we need to have here. For purposes here, what is important is that the assumption that some such principle is part of the otherwise plausible permissibility theory is what shows Exclusion Beta off *to its best advantage*. It shows that Exclusion Beta can tell us what is wrong with a2. And, since we are likely, at an intuitive level, to think that a2 *is* wrong, we beg no question against Exclusion Beta if we assume that the otherwise plausible permissibility theory will contain a principle that deals with tradeoffs in a way that directs that a2 (given the option of a3) is wrong.

2.7.9 So far, so good. But so far *we have just evaluated* a2. When we turn to a1, Exclusion Beta does not do as well. It creates the same obligations to do the same avoidable and unjustifiably bad things to the merely possible that Exclusion Alpha does in the case where w1 is actual. According to Exclusion Beta, Etta's losses *alone* matter morally for purposes of evaluating a1 since she alone exists at w1. That is: the losses Etta incurs at w1 and w3, and those losses alone, have moral significance for purposes of evaluating a1. Moreover, a2 makes things better for Etta than a1 does. And it does so without – according to Exclusion Beta – imposing any morally significant loss on anyone else at all. Thus Fen incurs a loss at w2 (compared to w3). But that loss has *no moral significance* for purposes of evaluating a1, according to Exclusion Beta, since Fen never exists at w1. Given these moral data, OPPP1 instructs that a1 is wrong. But that result seems false.

So Exclusion Beta takes us to the same counterintuitive result that Exclusion Alpha does, given the supposition that w1, not w2, is actual. Exclusion Beta errs, just as Exclusion Alpha does, when it blinds itself to the loss that Fen incurs at w2 for purposes of evaluating a1.

2.7.10 Now, as noted earlier in connection with Exclusion Alpha, one might always argue that intuition has gone awry and that a1 *is* wrong. I don't think a1 is wrong. But even if it is, we should at least agree that Fen's loss at w2 *bears on* the permissibility of a1. Even the Totalist, bound to conclude that a1 is wrong, agrees that Fen's loss at w2 *bears on* the evaluation of a1. But according to Exclusion Beta, it doesn't. And that's a problem.

2.7.11 Like Exclusion Alpha, Exclusion Beta is at the same time both *too exclusive* and *not exclusive enough*. In both Double Wrongful Life and Addition Plus, Exclusion Beta is *too exclusive*. It screens out losses that are clearly morally significant. Dee's loss under a2 surely bears on the permissibility of a1. In a roundabout way, Dee's loss counts in favor of a1 by showing that a2 itself is too flawed to show that a1 is wrong. Cal's plight under a1 just as surely bears on the permissibility of a2 in the same roundabout way.

And in Addition Plus Fen's loss at w2 clearly counts, not just *against* a2, but also, in a roundabout way, *in favor of* a1. Fen's loss helps explain why we think a1 is permissible – or, at least, why we hesitate before we declare it wrong.

At the same time, in the Basic Case, Exclusion Beta is *too inclusive*. It does what it is supposed to do when we evaluate a1. But when we turn to a2, Bob's loss at w1 abruptly assumes *full moral significance* – forcing any otherwise plausible permissibility theory just as abruptly, having already declared a1 *obligatory*, now to say that a2 is *permissible*.

The better view is that the loss incurred by Bob at w1 is devoid of moral significance *regardless* of the act we happen to be evaluating. As noted before, the trick is to say that but *not* say, in Addition Plus, that it is permissible to bring Fen into existence and treat him badly.

2.8 Inclusion

2.8.1 Exclusion Alpha and Exclusion Beta both comport with our intuitions only about half-way. This means that the fact that Inclusion does not – as we shall see – *perfectly* comport with our intuitions does not give us a reason to prefer Exclusion. Moreover, Exclusion in both its forms faces serious conceptual difficulties – difficulties that Inclusion has the resources to avoid nicely. Thus, Inclusion avoids the mistake of linking an act's permissibility to its performance, and never even hints that an *obligatory* act can have *permissible* alternatives.

These realities have very likely led many theorists to consider Inclusion a serious alternative to Exclusion and – mistakenly, I think – a plausible view in its own right.

Inclusion:

> For any person *p* who does or will exist at any world, any loss incurred at any world *w* by *p* – including the loss incurred by *p* when agents fail at *w* to bring *p* into an existence worth having – is morally significant for purposes of determining the permissibility of any act *a* performed at *w* that imposes that loss and any alternative act *a′* performed at any other accessible world *w′* that avoids that loss. That is so, regardless of whether *p* does or will exist at *w*.

According to Inclusion, *all* losses incurred by *all* persons at *all* worlds – including losses incurred at worlds where the persons who incur those losses never exist at all – matter morally for purposes of evaluating *all* acts, independent of who actually exists and independent of which act in fact happens to be under scrutiny. Inclusion thus casts a wide net on the morally significant loss.

2.8.2 Inclusion's intuitive and conceptual advantages are on display in Double Wrongful Life. Inclusion recognizes that Cal's loss at w1 is morally significant for purposes of determining the permissibility of a2. Accordingly, the otherwise plausible permissibility theory that is combined with Inclusion does not blind itself to that loss in determining the permissibility of a2. It can say, in particular, that the loss Cal incurs at w1 shows that a2 is really not any more dreadful than a1 is. The loss Cal incurs at w1, by *counting against* a1, in a roundabout way, thus counts *in favor of* a2. The way is thus paved for the result that a2 is permissible. By the same logic, a1 is permissible as well. Those results seem clearly right.

2.8.3 In Addition Plus, Inclusion considers the loss Fen incurs at w2 – the loss established, of course, by comparing Fen's wellbeing level at w2 against Fen's wellbeing level at *w3*, not *w1* – morally significant for purposes of evaluating a1. That is so, even if a1 is actually performed and Fen remains merely possible, and despite the fact that Fen would not exist if a1 were performed.

Now, we might quarrel about what a correct permissibility theory will do with that particular moral datum. Perhaps it will deem a1 wrong on the view that agents ought to maximize aggregate wellbeing, and consider a2 accordingly obligatory, despite the morally significant loss Fen incurs at w2. Or perhaps it will deem a1 wrong because it considers permissibility to be determined on the basis of *both* aggregate wellbeing *and* the distribution of wellbeing across the population. Taking that approach, we might say that a2 shows that a1 is wrong *and* that a3 shows that a2 is wrong but that a3 is itself permissible – that is, that there is nothing in a1 that shows that a3 is wrong. Or perhaps it will even find a way of saying that a1 is permissible or even obligatory, despite the morally significant loss Etta incurs at w1, on still another form of pluralism, a form that considers human flourishing a good, alongside aggregate wellbeing and equality.[75] What is most important for purposes here is that Inclusion *correctly* insists that the loss Fen incurs at w2 must be *taken into account* in evaluating a1; Fen's loss at w2 *bears on*, though does not on its own determine, whether a1 is permissible or not.

2.8.4 Inclusion avoids the conceptual difficulties that plague Exclusion Alpha and Exclusion Beta by making the class of losses deemed to have moral significance *fixed*. Double Wrongful Life and Addition Plus make that point. We encounter no surprises when we change suppositions regarding which act is actually performed and no inconsistencies when we shift from evaluating one act to evaluating another. Results remain stable, deontic axioms obeyed and inconsistencies avoided.

Moreover, by making the class of losses deemed to have moral significance *very large*, Inclusion avoids the intuitive difficulties Exclusion Alpha and Exclusion Beta create. Losses incurred by the merely possible – and those who would remain merely possible even had the act under scrutiny itself been performed – are understood to bear on permissibility. Double Wrongful Life and Addition Plus make that point as well. Thus Inclusion steadfastly recognizes that the loss Dee incurs at w2 bears on what is done at w1, the loss Cal incurs at w1 bears on what is done at w2 and the loss Fen incurs at w2 bears on what is done at w1. Intuitively, each of those losses does matter morally, not just in the sense that it counts against the act that imposes it, but also that it, in a roundabout way, counts in favor of the act that avoids it.

2.8.6 At the same time, the Basic Case shows that Inclusion is surely false. We intuitively think Ann must – in some way – come first. But that is not how Inclusion sees things. According to Inclusion, there is nothing to recommend a1 over a2. We can make people happy, we can make happy people. The otherwise plausible permissibility theory that comports with Inclusion will instruct that we may do as we please.

The Basic Case reveals something important about Inclusion: it puts the *merely possible in a wildly implausible competition with those who do or will exist.* The view that you have done something perfectly permissible when you refuse to rescue your own *real live dog* from a "terrible spat" for sake of the Calico Cat – or Thomasina or the Cat Who Went to Heaven or any other *merely possible cat* – seems just mistaken.

---

[75] See Temkin (1993).

In a way that can only be described as axiomatic, your actual dog *must* – in some way – come before any merely possible cat. We must find a way of saying that Ann comes before Bob even as we insist that Cal does not come before Dee (or vice versa) and that Etta does not come before Fen.

We must, in other words, find a way of saying (A) that Ann comes first *without* taking the position that (B) if w1 is actual in Addition Plus, and a1 is actually performed, then a1 is wrong. But we must also find a way of saying (A) without taking the position that (C) an act can be permissible, or even obligatory, if unperformed but wrong if performed. And finally we must find a way of saying (A) without taking the position that (D) an act is *obligatory* but that its alternative is – in the same case – somehow *permissible*. We must, in other words, avoid both Inclusion and Exclusion, in both of its forms.

2.9 Variabilism

2.9.1 A middle ground between Inclusion and Exclusion is Variabilism. The fact that Variabilism is at least plausible means that, having rejected the frying pan, we should not feel immediately compelled to leap into the fire.

Variabilism can be put as follows.

Variabilism:

A loss incurred at any world *w* by any person *p* has moral significance for purposes of determining the permissibility of any act *a* performed at *w* that imposes that loss and any alternative act *a′* performed at any accessible world *w′* that avoids that loss *if and only if p* does or will exist at *w*.

By implication, the loss incurred by *p* at *w* when agents fail to bring *p* into an existence worth having at *w* will have no moral significance for purposes of determining the permissibility of the act *a* performed at *w* that imposes that loss or any alternative act *a′* performed at *w′* that avoids that loss.

On this view, whether a loss bears on the permissibility of an act depends not on *who incurs that loss* but rather on *where that loss is incurred* in relation to the person who incurs it. For any particular person, some losses have moral significance for purposes of determining the permissibility of the act under scrutiny and its alternatives and some do not. A loss incurred by a person at a world is devoid of moral significance for purposes of evaluating the act that imposes that loss or any alternative to that act if that loss is incurred at a world where the person who incurs that loss never exists at all. Incurred, however, at a world where that person does or will exist, a loss will have full moral significance.

2.9.2 Variabilism functions in the Basic Case just as Exclusion Alpha does on the supposition that a1 is actually performed – and as Exclusion Beta does (given the inconsistency, perhaps we should say *tries* to do) when it is a1 that is under scrutiny. According to Variabilism, the loss Ann incurs at w2 bears on the permissibility of both a1 and a2 since Ann exists at w2. In contrast, the loss Bob incurs at w1 has no moral significance whatsoever, for purposes of evaluating either a1 or a2, since Bob never exists at w1. Given those moral data, the otherwise plausible permissibility theory will consider a1 obligatory and a2 wrong.

An advantage over Exclusion Alpha is that this account and these results will hold *whether or not it is w1 that is actual*. Suppose that w2 is actual and that a2 is actually performed instead of a1. The loss Bob incurs at w1 will remain, according to Variabilism, devoid of moral significance *despite the fact that Bob now actually exists*. We are thus never forced to say that the wrong act – that is, a2, the act that is wrong if left unperformed – becomes right if performed.

Variabilism has a similar advantage over Exclusion Beta. *Whether we happen to be evaluating a1 or a2*, the loss Bob incurs at w1 remains devoid of moral significance. Variabilism thus has no opportunity – provided, of course, that the permissibility theory is itself otherwise plausible – to generate *both* the result that a1 is obligatory *and* the result that a2 is permissible.

2.9.3 The conceptual advantage Variabilism has over Exclusion in both its forms is a function of the fact that, under Variabilism, the class of losses that have moral significance is *fixed*. The losses that bear on the permissibility of any one act or any alternative to that act do not shift depending on which act is actually performed or which act we happen to be evaluating. Variabilism thus achieves a certain stability, and consistency, across its various results.

Now, so does Inclusion. But Variabilism has clear advantages over Inclusion. Unlike Inclusion, Variabilism considers Bob's loss at w2 morally insignificant – and we accordingly avoid Inclusion's impression that a1 and a2 are nothing more than two peas in a moral pod. According to Variabilism, Ann *must* come before Bob; your actual dog *must* come before your merely possible cat. We thus find in Variabilism but not Inclusion the thesis that it *does* matter morally whether we *make people happy* or *make happy people*.

2.9.4 Variabilism also suggests a plausible account of Double Wrongful Life. According to Variabilism, the losses incurred by Cal at w1 and Dee at w2 *both* bear on the permissibility of *both* a1 and a2. If imposing on Dee the loss at w2 counts against a2, avoiding the loss Cal incurs at w1 counts, in a roundabout way, in favor of a2. If imposing on Cal the loss at w1 counts against a1, avoiding the loss Dee incurs at w2 counts, in a roundabout way, in favor of a1. Variabilism functions here just as Inclusion does – and parts ways with Exclusion. It appreciates a certain loss that Exclusion Alpha and Exclusion Beta blind themselves to – and can, accordingly, appreciate the parallels. The otherwise plausible permissibility theory will imply that a1 and a2 are both permissible.

2.9.5 Finally, we turn to Addition Plus. Let's first consider a1 – and compare it, first, against a2. According to Variabilism, the loss Etta incurs at w1 is morally significant. But so is the loss Fen incurs at w2 (compared to w3, of course, not w1). It's true that a2 avoids the morally significant loss Etta incurs at w1. But the morally significant loss Fen incurs at w2 is much deeper. Now – as noted before – just how this tradeoff between Fen and Etta is to be made is something we leave to the otherwise plausible permissibility theory. My own view is that a more complete version of that theory will say that the fact that a2 imposes the very great and very deep loss that it does on Fen – compared, again, not to how well off Fen is at w1, but rather to how well off Fen is at w3 – means that a2 is in no position to establish that a1 is wrong.

In any case, what is very clear is that Variabilism, by understanding Fen's loss at w2 to *bear on* the permissibility of a1, parts ways with Exclusion Alpha (in the case where a1 is actually performed) and Exclusion Beta (since a1 is the act under scrutiny).

But Variabilism also parts ways with Inclusion. While Inclusion counts the loss that Fen incurs at w1 *against* a1 – making it that much harder to find a1 permissible – Variabilism instead treats that particular loss *as though it never happened.*

To complete the evaluation of a1, we must also compare a1 against a3. (So far, all we have done is to establish that a2 does not show that a1 is wrong.) According to Variabilism, the loss Etta incurs at w3 is morally significant since Etta exists at w3. But what about the loss that Fen incurs at w1? Here again, Variabilism parts ways with Inclusion – instead tracking Exclusion Alpha (in the case where a1 is actually performed) and Exclusion Beta (since it is a1 we are evaluating). It's true that a3 avoids the loss that Fen incurs at w1 and that a3 makes things better for Fen than a1 does. But there's a catch. According to Variabilism, since Fen *never exists at all* at w1, the loss Fen incurs at w1 has no moral significance. Thus, what we really have in a3 is an act that imposes a morally *significant* loss on Etta while avoiding a morally *insignificant* loss on behalf of Fen. To top it all off, a1 makes things better for Etta than a3 does. On those facts, there is nothing in the otherwise plausible permissibility theory that even begins to suggest that a1 is wrong.

But this is not yet to say that a1 is *permissible*. OPPP2 does not apply since a1 imposes a morally significant loss on Etta. OPPP3 does not apply since the changes are not merely reversing. Now, plausibly, if the principles within our relevantly complete permissibility theory do not imply that an act is wrong, we should think that that act is permissible. But we don't yet have a relevantly complete permissibility theory; this kind of inference is, accordingly, premature.

Indirectly, however, we can produce a firmer foundation for the result that a1 is permissible. And that is to rule out a2 and a3 – to find them wrong. Now, we won't be able to do this by reference just to OPPP1–OPPP4; we will have to speculate a bit how an otherwise plausible permissibility theory will direct that one particular tradeoff is to be decided. But once we do that, we are then done. For at that point standard deontic axioms will instruct that, if a2 and a3 are wrong, a1 is permissible.

Accordingly, we turn to a2 and a3. But this discussion will be brief, for, according to Variabilism, what counts as a morally significant loss does not vary depending on the act under scrutiny. So what we have said before about which losses are morally significant and which are not holds for the present discussion as well.

Let's take a2 first – and compare it against a3. Fen alone incurs a loss at w2 and Etta alone incurs a loss at w3. According to Variabilism, both of those losses are morally significant. We, accordingly, face a tradeoff. It is here that we can only speculate regarding the contents (beyond OPPP1–OPPP4) of the otherwise plausible permissibility theory. It seems plausible, however, to think that the tradeoff should be made in favor of Fen rather than Etta. In other words, the fact that a3 makes things better for Fen than a2 does, and thereby avoids a morally significant loss on his behalf, all without imposing *too* deep a morally significant loss on Etta, plausibly means that a2 is wrong.

This analysis, reminiscent of Leximin, is somewhat controversial. Certainly, the Totalist and perhaps the Pluralist, while agreeing that Fen's loss at w2 is morally significant, will still find a2 obligatory. But a tradeoff principle that declares a2 wrong in light of the availability of a3 may seem more plausible.

We turn, finally, to a3. Here it's the comparison against a1 that is telling. a3 imposes a morally significant loss on Etta. a1 makes things better for Etta than a3 does. Now, it's true that Fen incurs losses at w1 (as compared to both w2 and w3). According to Variabilism, however, those losses have no moral significance since Fen never exists at w1. On these facts, the otherwise plausible permissibility theory – and, specifically, OPPP1 – will imply that a3 is wrong.

We conclude, then, that a2 and a3 are both wrong, and a1 is obligatory.

2.9.6 These are highly plausible, and consistent, results. We are seeing here the *variable moral significance* not just of *merely possible people* – Fen at w1 – but also of *existing or future people* – Fen at w2 and at w3. In some respects, the merely possible matter morally – just as, in some respects, do you and I. When we ask whether a2 shows that a1 is impermissible, what we see is that Fen, though merely possible relative to w1, matters morally just as much as Etta does. But in other respects, the merely possible do not matter at all – and nor do we. Thus, leaving Fen out of existence at w1 imposes, according to Variabilism, no morally significant loss on Fen. At that juncture of the analysis, Fen does not matter morally *at all*, not even the littlest bit, for purposes of evaluating a2 or a3 (or a1).

2.10 The Neutrality Intuition

2.10.1 Variabilism does not exhaust the middle ground between Exclusion and Inclusion. It shares that ground with some other views. One such view is suggested by what John Broome calls the *Neutrality Intuition*. Another is what Peter Singer calls the *Prior Existence View*. In certain respects, these two views are closely related. Each can be viewed as an attempt to capture our *person-affecting*, or *person-based*, intuitions. And each is in the end rejected – rightly, I think – by its respective articulator on the ground that it excludes from the analysis the interest of the person whose coming into existence is at stake.

2.10.2 According to the Neutrality Intuition, "adding a person [to the world] is equally as good as not adding her" – that is, "adding a person to the world is very often ethically neutral."[76] Broome considers the Neutrality Intuition "part of the broader way of thinking known as the 'person-affecting view.' This is the view that

[76] Broome (2004), pp. 143 and 146.

only benefits or harms that come to people can be ethically significant.... A change is neutral unless it makes someone either better or worse off than she would otherwise have been."[77]

Broome argues that we should reject the Neutrality Intuition. I agree. What I want to point out here is that his argument against the Neutrality Intuition does not apply to Variabilism. In fact, Variabilism provides a good account of just why it is that the Neutrality Intuition fails.

2.10.3 Broome's first step in constructing an argument against the Neutrality Intuition is to state the view in a more precise way. He proposes the *Principle of Equal Existence*, which pertains, not to *act* permissibility, but rather to *outcome* betterness.[78] I paraphrase it here:

> *Principle of Equal Existence*: If two outcomes, or *distributions*, include the same populations except that one contains an additional person whose life is in the "neutral range" – is, at least, not a life *less* than worth living – then those distributions are equally good *if and only if* they are equally good *for each person who does or will exist in both*.[79]

It is easy to see that the Principle of Equal Existence is an alternative to Exclusion and Inclusion. The Principle of Equal Existence does not require that people be actual in order to have their interests taken into account for purposes of ranking the distributions. The merely possible matter morally, too – as long as they exist in both the distributions to be ranked. At the same time, the principle excludes from the calculation the interests of the additional person even if that additional person is actual.

By implication, then, *leaving a person out of an existence worth having* is not going make one distribution worse than the other, nor is *bringing a new person into an existence worth having* going to make the other distribution better than the one.

---

[77] Broome (2004), p. 145. Interestingly, the motivation that Broome cites for the person-based approach is just that bringing a person into existence "makes her neither better nor worse off." That metaphysical point, he suggests, grounds the moral intuition for those who happen to have it. But we can also reject the metaphysical point – and I do; I think we can, and need to be able to, say that the person who never exists incurs a loss in the case where agents could have brought that same person into an existence worth having – and accept the person-based intuition on purely moral grounds.

[78] By *distribution*, Broome means to indicate a particular way in which wellbeing is distributed across a particular population. For purposes here, I have instead used to the term *outcome*. (A *world* is different: two worlds may differ in many details but still generate the same outcome, or distribution – that is, the identical distribution of wellbeing across the identical population. And we still need worlds: we can talk about acts performed at worlds producing certain outcomes; but acts themselves can produce vastly different outcomes, depending on the worlds at which those acts are performed.) So for purposes here either term will do. The more material distinction is between outcome *betterness* and act *permissibility*. This distinction I address in the text. See Broome (2004), pp. 140–149; see also note 7 above (distinction between *deontic* and *teleological* project) and Appendix C.

[79] Broome (2004), p. 146. The Principle of Equal Existence is formulated in terms of the "neutral range," and I assume here that the additional person whose life is "clearly worth living" has an overall wellbeing level that falls within that range. See also Broome (1991), p. 229 (discussion of the *Constituency Principle*).

2.10.4 As Broome shows, however, the Principle of Equal Existence leads to inconsistency when put together with still other principles we want to accept. Now, under the Principle of Equal Existence, it is *distributions* that are to be *ranked* in respect of their relative betterness, and not *acts* that are to be evaluated for their *permissibility*. With that caveat, we can display the sort of case that worries Broome in the usual way, with the alternative acts listed on the first row, and the distribution that is correlated in the case with each such act described below. As usual, the name in italics coupled with the asterisk means that the specified person never exists at all under the specified act, while boldface means that the specified person does, or will, exist under the specified act:

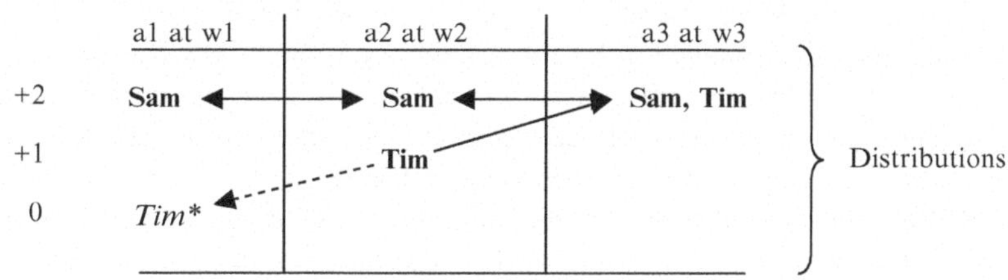

Case 2.4 Simple Addition

The Principle of Equal Existence implies that the distributions generated by a1 and a2 are equally good and so are the distributions generated by a1 and a3. The transitivity and symmetry of *equally as good as* implies that the distributions generated by a2 and a3 are also equally good. But we understand, on other grounds, that the distribution generated by a3 is clearly better than the distribution generated by a2.[80] So we have an inconsistency. We need to reject the Principle of Equal Existence.

2.10.5 Broome elsewhere suggests that coherent theories of act permissibility are easier to come by than are coherent theories of distribution goodness.[81] Interestingly, however, the root defect in the Principle of Equal Existence isn't at all mitigated if we extend that principle – and the idea behind it, that is, the Neutrality Intuition, the intuition Broome associates with the person-based approach – in a way that evaluates acts for their permissibility. Such an attempt will seem perfectly natural if we happen to accept – on the theory that the *consequences* of acts are critical for determining the *permissibility* of those acts – a strong connection between distribution goodness and act permissibility. What we will find, however, is that the results we obtain for act permissibility are just as problematic as they are for distribution goodness.

The case we've just examined – Simple Addition – makes this point. Suppose that a1 is itself permissible. Then, since the only difference between what a1 does and what a2 does is that a2 brings Tim into existence and accords to Tim an existence worth having, we should be able to infer that a2 is permissible as well. After

[80] Here Broome appeals to the *Principle of Personal Good*. Broome (2004), pp. 246–47.

[81] See Broome (2009), p. 248.

all, on the assumption that there is a strong connection between act permissibility and distribution goodness, if the distributions generated by a1 and a2 really are equally good, and if a1 is permissible, then a2 is permissible as well. For, if it were true that a1 is permissible and a2 wrong, then we should be able to see that the distribution generated by a2 is morally worse than the distribution generated by a1. But we don't, according to the Principle of Equal Existence. By the same analysis, a3 is permissible. But the fact that the distribution generated by a3 is *better than* the distribution generated by a2 should tell us that a2 is impermissible. And we again we face inconsistency: a2 can't be both *permissible* and *wrong*.

2.10.6 Broome's conclusion is that we must reject the Principle of Equal Existence. It doesn't work as a criterion for when one distribution is as good as another. Nor – as we have just seen – do things get any better when we extend the principle in a way that makes it relevant to the issue of act permissibility.

Variabilism provides us with a clear account of just why this principle and the Neutrality Intuition itself both fail. Let's stay, for the moment, on the topic of act permissibility. For that purpose, we need to retain the assumption that act permissibility and distribution goodness are strongly connected. Variability diverges from the extended Principle of Equal Existence when the latter claims that a1 and a2 are *equally good* and hence that, if a1 is permissible, so is a2.

For it's at that point that the latter principle makes the mistake of focusing exclusively on what is good *for Sam*, who exists under both a1 and a2 and fares just as well under one act as he does under the other. It's at that point any *correct* principle should take into account the fact that *Tim incurs a loss under a2*– not, of course, a loss compared to how Tim fares at *a1*, but rather a loss compared to how Tim fares at *a3*. It's at that point that any correct principle will discern that loss and its relevance to the moral analysis. The approach to act permissibility that an extended Principle of Equal Existence suggests, however, blinds us to that very loss – or at least to its moral significance.

In contrast, Variabilism implies that the loss that Tim incurs at a2 has *full moral significance*, for purposes of evaluating a2 and each of its alternatives. After all, it is a loss Tim incurs at a world, w2, at which Tim exists. The otherwise plausible permissibility theory – the most standard of Pareto principles, and as well as the more restricted OPPP1, in combination with Variabilism – will quickly instruct that, given a3, a2 is wrong.

2.10.7 Variabilism thus avoids the inconsistency that we derived when we tried to extend the Principle of Equal Existence in a way that makes it pertinent to act permissibility. Does an *extended Variabilism* – Variabilism extended, that is, in such a way that it addresses certain issues of distribution goodness – similarly avoid the inconsistency we derived from the original Principle of Equal Existence?

Again, retaining the assumption that act permissibility and outcome goodness are strongly connected, we can speculate as follows. If a2 is wrong and a1 is permissible – as Variabilism, in combination with the otherwise plausible permissibility theory, will instruct – then we want to say that it cannot also be the case that the distributions generated by a1 and a2 are equally good. We want to say that, from a moral point of view, the distribution generated by a2 must be *worse than* the distribution generated by a1 is. By a parallel analysis, we also can say that, since a2 is wrong and a3 is permissible, the distribution generated by a2 must be worse than the distribution generated by a3, and that, since a1 and a3 are both permissible, the distribution generated by a3 is exactly as good as the distribution generated by a1.

Intuitively, these results seem plausible. Moreover, there is no inconsistency.

We obtain the same results even if we don't rely on the results we obtain regarding permissibility for purposes of ranking the distributions in respect of their goodness. According to Variabilism, no one incurs any morally significant loss whatsoever under a1. In contrast, Tim incurs a fully morally significant loss under a2. Any otherwise plausible *ranking* theory may well instruct, on those facts, that the distribution that a2 generates – its consequence, that is – is morally worse than the distribution that a1 generates. Adding a new person – Tim – can thus be plausibly be considered to have *made things worse* from a moral point of view.

According to the Neutrality Intuition, "adding a person to the world is very often ethically neutral."[82] But what the cases seem to suggest, and a view that Variabilism itself clearly supports, is that adding a new person to the world is very often *ethically hazardous*. This is not at all to say we shouldn't do it – in, for example, Simple Addition. It's just to say that if we do add a new person to the world, then we should do it by way of a3 and *not a2*. Who could disagree?

2.10.8 The Neutrality Intuition was proposed by Broome as a part of what he calls the person-based view.[83] Variabilism suggests an alternative understanding of the person-based intuition. We can use Simple Addition to illustrate the difference. The otherwise plausible permissibility theory itself – specifically, OPPP2 – implies that a2 is wrong *only if* a2 imposes a morally significant loss on some person or another. But this necessary condition on wrongdoing is consistent with an extremely wide person-based approach. It's consistent, for example, with Totalism. We don't get the *intuition* in the person-based intuition until we add a loss rule of a certain kind. Variabilism will certainly do for that purpose. Variabilism, in combination with OPPP2, thus implies that a2 is wrong at w2 *only if* a2 imposes a loss at w2 on a person who does or will exist at w2. And it seems that we should also be able to say that the distribution generated by a2 is *worse than* the distribution generated by a1 *only if* a2 imposes a loss at w2 on a person who does or will exist at w2. Thus:

> a2 performed at w2 is wrong, and the distribution generated by a2 performed at w2 is worse than the distribution generated by a1 performed at w1, *only if* a2 imposes a loss at w2 on a person who does or will exist at w2;

What we have here is a simple necessary condition on wrongdoing: a way of understanding the intuition that the "bad" act must be "bad for" someone or another.[84] But it's a necessary condition that can be satisfied by Tim – for example – in the

---

[82] Broome (2004), p. 143.

[83] Broome (2004), p. 145.

[84] Parfit (1987), p. 363.

context of Simple Addition – since Tim does incur a loss at w2. Thus: while Tim incurs no loss under a2 relative to a1, Tim clearly does incur a loss under a2 relative to a3. And that loss counts, for purposes of satisfying the necessary condition. The implication to the "no wrong done" result is, in other words, blocked.

In contrast, the necessary condition that the Neutrality Intuition establishes is more stringent. "A change is neutral unless it makes someone either better or worse off than she would otherwise have been."[85] Thus, where *unless* means *if not*, and where we take the contrapositive, the Neutrality Intuition states that the change from a1 at w1 to a2 at w2 can make things worse *only if* someone – Tim, for example – is worse off at w2 than that person – Tim – is at w1.

> The distribution generated by a2 performed at w2 is worse than the distribution generated by a1 performed at w1 *only if* a2 makes things worse for a person who does or will exist at w2 than a1 makes things for that same person who does or will exist at w1.

The Principle of Equal Existence makes that same point. But in the context of Simple Addition, this condition is not satisfied. Tim is, if anything, *better off* at w2 then he is at w1. And Tim in any case exists only at w2. And so we infer that the change from a1 to a2 can't make things worse. That is: a2 is not worse than a1. And that, of course, is just the inference that created the problem we noted above to begin with. But it's a problem that Variabilism, in combination with the otherwise plausible permissibility theory, nicely avoids, since the focus for Variabilism is *not* whether the existing or future Tim is worse off at w2 *relative to w1*, but rather whether Tim is worse off at w2 *relative to any other scenario whatsoever*.

One other note. In still another respect, the Neutrality Intuition is much broader than what we have in Variabilism in combination with the otherwise plausible permissibility theory. For the Neutrality Intuition also implies this:

> The distribution generated by a2 performed at w2 is better than the distribution generated by a1 performed at w1 *only if* a2 makes things better for a person who does or will exist at w2 than a1 makes things for that same person who does or will exist at w1.

The Principle of Equal Existence makes that same point. But this can't be right. Given that it isn't better for *anyone* that a1 be performed rather than a2, these principles imply that a1 is no better than a2 is. And given that we know that a2 is wrong (a3 shows us that that is so), and a strong connection between distribution betterness and act permissibility, we must now say that a1 is wrong as well. Now, this result may not seem immediately objectionable in connection with Simple Addition. But precisely the same line of reasoning applies in the case of Addition Plus as well – forcing us now to say that a1 in that case is wrong: that is wrong *not* to bring additional merely possible persons into existence (e.g. Fen) and *treat them badly*. But we understand that that result cannot be right.

What we are seeing here is that it is a mistake to think that the simple *necessary* condition that the person-based intuition is meant to express can be swollen to try to express the necessary *and sufficient* condition that Broome associates with the

[85] Broome (2004), p. 145.

"person-affecting view." The simple *necessary* condition on wrongdoing (understood by reference to Variabilism, not the Neutrality Intuition) is fine. But to convert that necessary condition on wrongdoing – on when one act is worse than another – into a *sufficient* condition is a *mistake*. In fact, it's a logical error – something close to the fallacy of affirming the consequent – to turn this simple necessary condition around, add an essentially innocuous condition and then consider it something we are *compelled* to accept should we have happened already to have committed ourselves to the simple necessary condition.

Logic itself, then, does not require that move, and it is one that we should, moreover, avoid. It is easy to why that is so, if we now shift the case, from Simple Addition back to Addition Plus. We should reject – and logically are perfectly free to reject, even if we accept the *necessary* condition on wrongdoing – the following *sufficient* condition on wrongdoing:

> a1 performed at w1 is wrong, and the distribution generated by a1 performed at w1 is worse than the distribution generated by a2 performed at w2, *if* a1 imposes a loss at w1 on a person who does or will exist at w1 and each person who does or will exist under a2 at w2 is at least as well off as that same person is under a1 at w1.

Addition Plus is an excellent counterexample against this variety of sufficient condition on wrongdoing. We think a1 is perfectly permissible, despite its effect on Etta. And – as we have already seen – it is a counterexample that Variabilism, in combination with the otherwise plausible permissibility theory, avoids.[86]

[86] For purposes here, I have assumed a strong connection between outcome betterness and act permissibility, and have freely gone back and forth from the one topic to the other. This approach clearly implies that the distributions generated by two acts can be ranked in respect of their betterness in one way if a third act producing still a third distribution is available to the agents, and in another way if it is not. The Axiom of the Independence of Irrelevant Alternatives could be taken to suggest that that cannot happen.

There are two points to be made in this connection. One is that there is no reason to think that the Independence Axiom does not apply nicely to the four-place betterness-for relation (the distribution generated by a1 at w1 is better for *p* than the distribution generated by a2 at w2 is for *q*) if not the two-place betterness relation (a1 is better than a2 *simpliciter*). The second is that, if we do think the Independence Axiom applies to the two-place relation as well, then it is plausible to think that what decides whether some a3 is an alternative for agents or not is in fact something inherent in a1 and a2. Thus: we might think natural law means that the world at which I am instantly cured of some terrible disease is not a world that is accessible to agents. But if that is so, then that is so because the world where I am sick keeps its agents firmly under the governance of that particular natural law. Taking that approach, when we specify (for purposes of determining permissibility) that a3 is an alternative to a1 and a2 for agents, and when we specify that a3 is not an alternative to a1 and a2 for agents, what we refer to by *a1* and *a2* itself shifts. In one case, there is something in the pair of options that precludes a3, and in the other case there isn't. We can thus retain the Independence Axiom for the two-place betterness relation as well as the four-place betterness-for relation. I have discussed these issues elsewhere. See Roberts (1998) and Roberts (2003b). See also Appendix C below.

2.11 The Prior Existence View

2.11.1 Though a theory of act permissibility and not distribution betterness, what Singer calls the *Prior Existence View* is similar to the Neutrality Intuition in certain respects. The Prior Existence View states that, when evaluating whether to bring a new person into existence, we are to "count only beings who already exist, prior to the decision we are taking, or at least will exist independently of that decision."[87] On this principle, we omit from our calculations the wellbeing the new person will accrue if the choice to bring that person into existence is made. That move, in turn, allows us to reject the highly counterintuitive idea that "the couple [should] count the likely future pleasure of their children as significant reason for having children…."[88]

2.11.2 As noted, Singer rejects the Prior Existence View. To see why, we need just consider the child "who, perhaps because it will inherit a genetic defect, would lead a thoroughly miserable life and die before its second birthday…." [89] Now, on some theories it will make a difference to the analysis just how miserable this "thoroughly miserable life" in fact is. The discussion can become very complicated in the case where the child's life is, though miserable, unambiguously worth having.[90] But those complications are not important here. For purposes here, we can simplify the discussion by stipulating that the child's life is *less* than worth living – that is, genuinely *wrongful* – if he or she exists at all.

According to the Prior Existence View, the fact that the child's life will unavoidably be less than worth living must be excluded from the moral evaluation of the choice whether to bring that child into existence or not. But surely we can't plausibly include that fact. "We would think it wrong for a couple

One last point on the Independence Axiom. If the approach I describe in this note turns out to be unworkable, then one option we have is to understand that the two-place betterness relation has no independent role to play in our moral analysis. We are very close to that result in any case. That is: the otherwise plausible permissibility theory relies, not on the two-place relation, but rather the four-place relation, to make its permissibility assessments. We can easily just understand the two-place relation as a short-hand way of saying that a particular act is wrong. To say, in other words, than an act a is better than an act a' – or, more precisely, that the distributions accordingly generated by a and a' rank in that particular way – is just another way of saying that a' is wrong.

See also Appendix C for discussion of an objection from Broome that targets, not the Neutrality Intuition as a way of articulating the person-based approach, but rather the person-based approach in combination with the idea that act permissibility and outcome betterness are themselves closely linked.

[87] Singer (1999), p. 103.

[88] Singer (1999), p. 103. Singer himself may anticipate that that calculation would be made on an aggregative basis if we accept the Prior Existence View. After all, Singer considers the Prior Existence View as an alternative to Totalism, whose aggregative feature Singer offers no objection to in this context.

[89] Singer (1999), p. 104.

[90] At that point, issues involving the nonidentity problem are triggered. We thus put that particular version of the case to the side for purposes here. But see part 2.2.5 above and Appendix B.

knowingly to conceive such a child...."[91] At the very least, the fact that a person's life will unavoidably be less than worth living *bears on* the permissibility of the choice to bring that person into existence to begin with. The Prior Existence View asserts that the one has nothing to do with the other. We should, accordingly, reject the Prior Existence View.

2.11.3 It may initially seem plausible that the one child's "likely future pleasure" does not count *in favor of* the choice to bring that child into existence. But it is also very clear that the distinct child's expected misery *counts against* the choice to bring that child into existence. But if the principle that justifies the first claim for us is that the person whose coming into existence is at stake does not *count* – if the principle that justifies the first claim for us is just the Prior Existence View – then we must not just accept the first claim but also reject the second claim. And that latter we cannot do.

Accordingly, Singer rejects the Prior Existence View. That much seems exactly right. On that basis, he then seems to accept a basic *symmetry* between happiness and misery.[92] That much is at least understandable. As we have just seen, we do consider the *misery* that a new person will predictably suffer to bear on the permissibility of the choice whether to bring that person into existence. So it seems that we must consider the *happiness* that the happy person can be expected to accrue to bear on permissibility as well. Misery counts against the one act under scrutiny, and happiness in favor of the other.[93]

Now, as a general, all-other-things-are-equal sort of proposition, that seems right. But we can discern in the pair of cases that we have at hand here a distinction that has nothing to do with that general proposition. When we say that the choice to leave the happy person out of existence is permissible, the happiness that is foregone under that choice is foregone by a person *who never exists at all*. And when we say that the choice to bring the miserable person into existence is clearly wrong, the misery that is incurred under that choice is misery that is incurred by a person *who does or will exist*.

With this distinction in mind, we can then use Variabilism to explain why happiness does not count in favor of the one act but misery does count against the other act. (We can explain, that is, what is called the *Asymmetry*.) It's not that the interests of the new person are to be excluded across the board from our calculation of the permissibility of the choice to bring that person into existence. Rather, that person's interests should be included, but on a *variable* basis. According to Variabilism, the loss the person incurs when that person is left out of an existence worth having has *no moral significance at all*. In contrast, according to Variabilism, the loss the person incurs when that person is brought into the miserable existence has *full moral significance*. The relevant distinction is not that one case involves happiness and the other misery, but rather that in one case the loss is incurred at a world where the

---

[91] Singer (1999), p. 104.

[92] Singer (1999), p. 104.

[93] See also McMahan (2009), pp. 54–60.

person never exists at all, and in the other the loss is incurred at a world where the person does or will exist.

We can thus (A) reject, with Singer, the Prior Existence View. We can also (B) accept as a general, all-other-things-are-equal sort of proposition – that is, subject to (C) following – that, just as creating additional happiness for a person bears on the issue of permissibility, so does creating additional misery for a person. But we can also (C) limit that general proposition in the way that Variabilism suggests. We can, that is, insist that (B) itself is subject to the principle that the loss that is implied whenever agents create less wellbeing for a person when they could have created more has *moral significance* for purposes of evaluating the choice that imposes that loss or any alternatives to that choice *if and only if* the person who incurs that loss does or will exist at the world at which that loss is incurred.

2.11.4 According to Variabilism, then, the Prior Existence View fails because it excludes from the permissibility analysis too many of the losses incurred by the person whose existence depends on how the choice under scrutiny is made. The same point holds for the Principle of Equal Existence – and, more generally, the Neutrality Intuition. Those principles fail as well because they exclude from the betterness analysis too many of the losses incurred by the person who exists under one but not the other of the two choices being ranked in accordance with their betterness.

In contrast, Variabilism insists that those losses do bear on the issue of permissibility – and betterness. It recognizes those very losses as morally significant for purposes of evaluating the choices that impose those losses and the alternatives to those choices that avoid those losses. Yet Variabilism recognizes those losses as morally significant without forcing us to jettison our intuition that the loss incurred by the person at a world where that person *never exists at all* is itself devoid of moral significance. We can still say that avoiding *that* loss, on behalf of that person, does not give us a reason to bring that person into existence. We can still insist, appealing to Variabilism, that such a loss *cannot* make the otherwise permissible choice that imposes that loss wrong.

2.12 The Asymmetry

2.12.1 Singer's discussion of the Prior Existence View takes him into the very deep waters of what is called the *Asymmetry*. Singer uses one-half of the Asymmetry to motivate the Prior Existence View – though feels forced by the other half of the Asymmetry in the end to reject the Prior Existence View.[94] Moreover, the Asymmetry is an important case in its own right – and an excellent test case for Variabilism.[95] So it is worth focusing on a bit more closely here.

[94] Singer (1999), pp. 103–105.

[95] See also Jeff McMahan, who gave a trenchant early description of the Asymmetry. McMahan (1981), p. 100, and Singer (1976), p. 92.

The Asymmetry comprises the following two claims:

The Asymmetry. Other things being equal, it would be wrong to bring a child into existence who would have a life less than one worth living, but it is permissible *not* to bring a child into existence even if that child would have a happy or even a wonderful life.

Singer and McMahan, as well as Ingmar Persson, recognize the commonsense appeal of the Asymmetry. But they all find it difficult to maintain. And each concludes, in the end, that we have no choice but to consider the additional person's prospective happiness – in my terms, the *loss* that person will incur if that person is left out of existence altogether – as morally significant for purposes of evaluating the choice whether to bring that person into existence or not.[96] They each in the end conclude that the fact that the loss incurred by the merely possible person when that person is left out of existence altogether can make the otherwise permissible act that imposes that loss wrong. We are left to think that, after all, as Singer puts it, the "likely future pleasure" of one's children is a "significant reason" for having them.[97]

2.12.2 We begin by graphing the pair of cases that make up the Asymmetry. Where m is the miserable child and h is the happy child, and the italicized name together with the asterisk signifies that the specified person never exists at all at the specified world and the boldface name signifies that the specified person does or will exist, the Asymmetry can be diagrammed as follows.

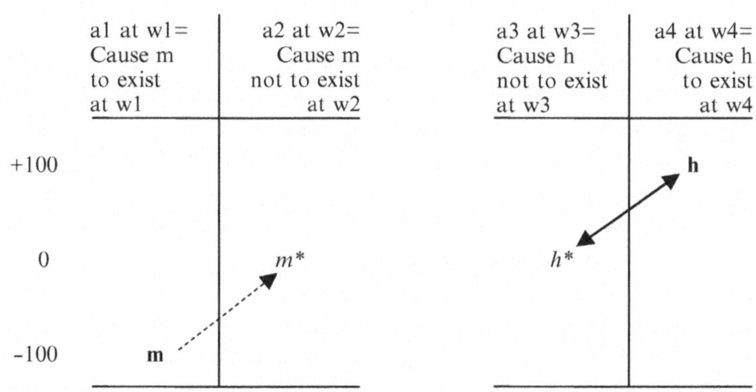

Case 2.5 The Asymmetry

It is the result that a1 is wrong but a3 is perfectly permissible – or, equivalently, that a2 is obligatory but a4 is not – that has seemed hard to explain. If a2 is something that we *must* do, then why isn't a4 also something we *must* do? If a1 is something we *can't* do, why is a3 perfectly permissible? If we have a strong moral reason for

[96] Singer (1999), p. 103. See also Persson (2009) and McMahan (2009).

[97] Singer (1999), p. 104.

leaving m out of existence altogether, why don't we also have a strong moral reason for bringing h into existence?

The problem is that many distinctions between a1 and a3, and between a2 and a4, seem incidental to the permissibility of those acts. As noted earlier, as a *general* proposition, we do think that just as misery can bear on permissibility, so can happiness. The idea that miserable children matter morally – and that accordingly any losses they incur have moral significance – but that happy children do not – and that accordingly their losses are somehow devoid of moral significance – is a non-starter. Moreover, the idea that losses below the zero level have moral significance whereas losses at or above the zero level do not seems arbitrary.

2.12.3 Exclusion Alpha and Exclusion Beta, of course, propose their own accounts in support of the Asymmetry. And the distinctions they rely on between a2 and a4 do intuitively seem to have *something* to do with why a2 is obligatory and a4 is not.

But those views come to the Asymmetry with their usual deficiencies fully intact. First, they only support our intuitions about half-way. Suppose a2 is actually performed. According to Exclusion Alpha, the loss m incurs at w1 has no moral significance at all. The otherwise plausible permissibility theory – specifically, OPPP2 – will then imply that a1 is permissible. Just as troubling, if a4 is actually performed, then Exclusion Alpha will say that the loss h incurs at w3 has full moral significance – and that a3 is wrong and a4, accordingly, obligatory. But both of these results are exactly contrary to what the Asymmetry says.

Exclusion Beta generates similarly problematic results whether it is a2 or a4 we happen to be evaluating.

Moreover, conceptual difficulties continue to plague both Exclusion Alpha and Exclusion Beta. Exclusion Alpha, in combination with the otherwise plausible permissibility theory, instructs that a1 is permissible if left unperformed – if, that is, a2 is actually performed – but instantly wrong if actually performed. a3 is permissible if actually performed – and instantly wrong if left unperformed. Exclusion Beta, still worse, instructs *for the same case* that a1 is wrong *and* a1 is permissible – and *for the same case* that a3 is permissible *and* a4 obligatory.

Inclusion simply denies the Asymmetry from the start.

2.12.4 Variabilism suggests a more plausible account of the Asymmetry. According to Variabilism, we do not decide whether a given loss has moral significance or not by tying that loss to a person and then trying to figure out whether that person matters morally or not – *counts* or not – for purposes of our evaluation. Instead, we recognize that *persons* themselves all matter morally in exactly the same way – in a way that is variable. It's their *losses* that do not all matter morally in exactly the same way. Some have moral significance, and some do not.

We need to consider, in particular, *where* the loss is incurred in relation to the person who incurs it. According to Variabilism, the loss incurred by either child m or h at a given world *w* bears on the permissibility of the act that imposes that loss and each alternative to that act that avoids that loss *if and only if* that child does or will exist at *w*.

Applying this loss rule, we divide the losses as follows. The loss m incurs at w1 has *full moral significance* for purposes of evaluating a1 and a2 since m exists at w1.

In contrast, the loss h incurs at w3 has *no moral significance whatsoever – not even the littlest bit* – for purposes of evaluating a3 or a4 since h never exists at all at w3.

It seems clear what the otherwise plausible permissibility theory will do with these moral data. On these data, any otherwise plausible permissibility theory – and, specifically, OPPP1 – will imply that a1 is wrong. The vast and morally significant loss m incurs at w1, in other words, where m does or will exist, makes a1 wrong, in light of the availability of the alternative a2, which is better for m and imposes no morally significant loss on anyone at all.

The same underlying analysis – specifically, OPPP2 – generates the result that a2 is permissible. Deontic axioms tell us still more – that, since a1 is wrong, a2 is in fact obligatory.

In dealing with the Asymmetry, the harder part has been to say what we have just said and *also say why a4 is not obligatory when a2 is* – that is, *why a3 is permissible*. According to Variabilism, however, what pushes us to say that a2 is obligatory – the *morally significant* loss that m incurs under a1 – is missing in the case of a4. For the loss that h incurs at w3, according to Variabilism, has *no moral significance whatsoever*. There thus *is* no morally significant loss that a4 avoids on behalf of h. The otherwise plausible permissibility theory, accordingly, implies that a3 is permissible – and so is a4. OPPP2 generates both those results. Variabilism thus agrees that a4 makes things better for h than a3 does; it agrees that h incurs a loss under a3 that a4 avoids on behalf of h. Variabilism simply asserts that that loss is devoid of moral significance (it is as though it never happened).[98]

Variabilism thus explains why the choice not to bring the happy child into existence is permissible while the choice to bring the miserable child into existence would be wrong. The difference is just that the loss for the miserable child takes place at a world where that real, live, flesh and blood child exists and suffers, whereas the loss for the happy child takes place at a world where that child never exists at all.

I suppose one might press harder on these two points. One might ask, for example, *why* the one loss matters morally and the other does not. But I think that we have said as much as there is to be said in response to that question. *It just does*. It's the difference between torturing your own real live dog and never bringing the Calico Cat or Thomasina or the Cat Who Went to Heaven into existence to begin with.

2.12.5 One last note in connection with the Asymmetry. The results that Totalism – as a paradigm example of the Inclusive approach – generates, regarding the first half of the Asymmetry, the half involving the miserable child, seem exactly right. But Totalism seems to go awry when applied to the second half of the Asymmetry, the half involving the happy child. It is there that we see that Totalism on the face of things seems to imply an implausibly stringent procreation obligation. We are

[98] These results parallel the results generated by the rights-based account that Persson explores. See Persson (2009). Persson himself rejects that account because of the difficulties inherent in the theory of rights that he is working with. I believe, however, that Variabilism, which is not rights-based, can track the plausible treatment of the Asymmetry that the rights-based account suggests while avoiding the difficulties inherent in that account.

alarmed! The Prior Existence View – which is a bit more exclusive – then comes to the fore. And the results that it generates for the second half of the Asymmetry seem exactly right. But then we apply it to the first half of the Asymmetry – and find that the results that it generates on that side are a disaster.

Variabilism arranges things so that we can take what we think is exactly right in a principled way from both Totalism and the Prior Existence View – and leave the rest. It fully supports the result that we do not have the stringent procreation obligation that Totalism itself suggests. That is, Variabilism, in combination with the otherwise plausible permissibility theory, flatly denies that we are required to perform a4. But it also flatly denies that that means that we can do whatever we want when it comes to bringing new persons into existence. Thus, in combination with any otherwise plausible permissibility theory, Variabilism insists that we are obligated not to bring the miserable child into existence – we are obligated, that is, *not* to perform a1. These are plausible results.

2.13 Summing Up

I thus believe Variabilism generates a plausible account of a significant range of problem cases. The loss incurred by the person who never exists at all cannot, according to Variabilism, make the otherwise permissible act that imposes that loss *wrong*. The accounts Variabilism provides of the Basic Case and the Asymmetry make that point. The implication for early abortion, and for non-conception and for conception and non-fruition, is clear: the option of an additional person coming into an existence worth having will not make the otherwise permissible choice to leave that person out of existence altogether wrong.

Just as important, Variabilism can say that without committing itself to the further result that what we have done is wrong in cases where *all we have done* is *decline* to bring a merely possible person into an avoidably and unjustifiably bad existence. That our choice *avoids* the maltreatment of the merely possible *can*, in other words, make the otherwise wrong act *perfectly permissible*. Double Wrongful Life, Addition Plus, Simple Addition and, again, the Asymmetry make that point.[99]

[99] I have proposed, in OPPP1 – OPPP4, some of the content of the otherwise plausible permissibility theory – which itself will consist of detailed principles that instruct, on the basis of the moral data that the loss rules under scrutiny will generate, which acts are permissible and which are not. We need, of course, to articulate – and test – more of that content. Appendix A (Otherwise Plausible Permissibility Theory + Variabilism) sums up what we have so far and suggests an extension.

Chapter 3
The Abortion Paradox

3.1 Introduction

Jeff McMahan recently has argued that a comparison between the choice of abortion and the choice of prenatal injury followed by fetal survival, in combination with certain Pareto-inspired principles, leads to an interesting paradox. I will call it the *Abortion Paradox*.

It seems that McMahan's own thought is that the Abortion Paradox arises in the case of late but not early abortion.[100] My take, however, is that the paradox is more interesting when we understand it to arise in the context of early abortion. That is, accordingly, how I will construct the paradox here. What is of particular interest here is the half of the paradox that argues *against* early abortion in some scenarios (Case 3.1).

The kernel of the Abortion Paradox is this. We assume that the child whose coming into existence is at stake will have an existence that is unambiguously worth having if that child exists at all. Then, it seems that *it is better for the child* that (i) an injury is inflicted on the early fetus – the fetus that will eventually develop into that child if there is no early abortion – and the fetus allowed to survive than that (ii) the early abortion take place. The choice of fetal injury is better for the child than the early abortion choice is. But – given that the agent also has a third option, the option of (iii) not injuring the fetus and not having the early abortion – we know that the fetal injury choice is wrong. But if that is so, how can the early abortion choice be *permissible*? How, more generally, can the *wrong* choice be *morally better than* the *right* choice?

In part 3.2, I consider, in a general way, the various Pareto principles that may be at play in the Abortion Paradox and how they relate to each other. In part 3.3, I explain why I think one such principle – *Pareto Plus* – in particular should be a

[100] The text of McMahan's paper does not entirely resolve the issue. According to McMahan, neither early nor late abortion ends the life of what he considers to be a "person." Yet in both cases the fetus itself seems to have certain "time-relative interests." McMahan (2006), pp. 627–628. In subsequent informal conversation, however, McMahan has stated that his intent in the paper was to address issues relating to late abortion.

M.A. Roberts, *Abortion and the Moral Significance of Merely Possible Persons: Finding Middle Ground in Hard Cases*, Philosophy and Medicine 107,
DOI 10.1007/978-90-481-3792-3_3, © Springer Science+Business Media B.V. 2010

matter of concern to us. In part 3.4, I state the Abortion Paradox in a more detailed way. (Some readers may want to skip parts 3.2 and 3.3 go directly to part 3.4!) In part 3.5, I discuss the solution to the Abortion Paradox that McMahan himself proposes. In part 3.6, I propose a Variabilist alternative to McMahan's solution. That solution rejects Pareto Plus and on that basis rejects the Abortion Paradox. Part 3.7 says more about Variabilism and Pareto Plus, and part 3.8 about the relation among the Standard Pareto Principle, Pareto Plus and OPPP1 – which last could aptly be called *Pareto Minus*.

3.2 Pareto Principles

3.2.1 Variabilism generates data regarding the moral significance of various losses that people incur. It does not generate any permissibility results at all in the absence of an otherwise plausible permissibility theory. As a start in that direction, I proposed principles OPPP1–OPPP4 in part 2.2.4 above.

One of those principles – OPPP1 – was a Pareto principle.

Otherwise Plausible Permissibility Principle 1 (*OPPP1*):
An act a performed at a world w is wrong *if*:

(i) Act a imposes a morally significant loss on a person p at w;
(ii) An act a' at an accessible world w' performed in lieu of a at w is better for p at w' than a is for p at w; and
(iii) Either (a) a' imposes no morally significant loss on any person q at w' other than perhaps p, or (b) if a' imposes a morally significant loss on a person q at w', then a imposes a morally significant loss on q at w, and a' is at least as good for q at w' as a is for q at w.

In Chapter 2, we saw that Variabilism, in combination with OPPP1–OPPP4 and various deontic axioms, yielded plausible results in several cases while avoiding the intuitive and conceptual problems that plague Exclusion (in both its Alpha and Beta forms) and Inclusion.

A main goal of this Chapter 3 is to examine two other Pareto principles. The first principle – what I will call the *Standard Pareto Principle* – is restricted to cases where exactly the same people do or will exist under the acts that are to be compared. It is restricted, that is, to *same-people* cases – or, since he thinks individuals other than persons have moral status, what McMahan might call *same-individual* cases. The Standard Pareto Principle is intuitively compelling and stands up to rigorous testing.

But it seems that its restrictions mean that the Standard Pareto Principle will be of little relevance to the Abortion Paradox. For the cases of interest here have to do with our obligations in respect of bringing certain *additional persons into existence*.[101] Now, still another Pareto principle is not so heavily restricted. It's *meant* to address *additional-people* cases. Borrowing from Partha Dasgupta, I will call it *Pareto Plus*.[102]

3.2.2 Pareto Plus is of interest here because of the critical role it seems to play in a certain argument against early abortion. That argument is itself exactly half of McMahan's Abortion Paradox.[103]

The Abortion Paradox doesn't argue that *all* early abortions are wrong. It argues that some are wrong and others are permissible. And it argues that that is so, even in case where the scenario in which we determine that the early abortion is wrong appears to have all of its morally relevant features in common with the scenario in which we determine that the early abortion is permissible. And that is the paradox.

Now, in constructing the paradox, McMahan does not *explicitly* invoke Pareto Plus. Moreover, the metaphysics that he adopts – in particular, the claim that the individual that is the fetus is identical to the individual that eventually becomes (after birth, in McMahan's view) a person – may make us think that his own view is that the half of the Abortion Paradox that argues against early abortion relies just on the perfectly innocuous Standard Pareto Principle and not on Pareto Plus.[104]

---

[101] McMahan may deny that the cases that constitute the Abortion Paradox are in fact additional-person cases. But see note 143 below.

[102] Suppose that "the only relevant difference between X and Y … is that Y would have an additional person enjoying a positive level of well-being. Call the conception that says that therefore Y is the better world the 'Pareto-plus Principle.'" Dasgupta (1994), pp. 382–383. Now, in one respect, Dasgupta's principle is stronger than the principle I call Pareto Plus here. That is, in the context of McMahan's construction of the Abortion Paradox, the reason that the one act is said to be morally better than the other is – as we shall see – *not* that the one act adds a person who has an existence worth having. Consistent with the principle I call Pareto Plus here, that kind of addition can be considered a *neutral* change – a change that makes things neither better nor worse. The reason that the one act is said to be morally better than the other is rather that the *fetus*, which exists under *both* acts, has *some* moral status (according to McMahan) and is better off under the one act than it is under the other.

Interestingly, so long as we accept Pareto Plus, the Abortion Paradox goes through even if we jettison McMahan's view that the fetus itself has some moral status. For if we accept Pareto Plus, we should also accept that when the change really is a *mere* addition – when all persons who do or will exist under one act do or will exist under the other as well, and vice versa, and are exactly as well off under the one act as they are under the other, and vice versa – then the addition of a person whose life is worth living at least does not make the one act worse than the other.

The principle I call Pareto Plus here, while weaker that Dasgupta's, is still notable, and not just in respect of the role it plays in the Abortion Paradox. For example, it is all that we need for purposes of constructing the *Mere Addition Paradox*. For it yields the – apparently – modest claim that the mere addition of a person whose existence is worth having does not *make things worse*, and that claim is, as we shall see, the heart of the Mere Addition Paradox.

[103] McMahan (2006), pp. 625–655.

[104] But a different metaphysics is not going to make the substantive moral debate go away. See note 143 below.

For these reasons, precisely how his argument in fact progresses is a question of interpretation. But there is also some question – as we shall see – whether his argument can be effectively launched *without* Pareto Plus. Moreover, there is no question at all but that, whatever the better interpretation of McMahan, Pareto Plus enables us to construct a *very tidy argument against early abortion* – and an argument that the defender of early abortion must counter.

3.2.3 The twin scenarios McMahan describes as the basis for the paradox are not *identical* twins. They have in common that exactly three options are made available to the agent: (i) early abortion, (ii) non-lethal fetal injury and (iii) no abortion and no fetal injury. But they differ in the *order in which they are presented to and deliberated upon by the agent*. They differ in respect of their *order of presentation*.

That factual distinction is critical to the solution to the Abortion Paradox that McMahan himself proposes. It explains, he argues, why the early abortion is permissible in one scenario but not the other. When the order of presentation varies, so does the line of reasoned moral analysis triggered within the agent's own mind. McMahan's suggestion, then, is that if we believe that the reasoning that goes into a particular choice helps to determine the permissibility of that choice, we can resolve the paradox.

Many theorists *believe* that the permissibility of an act cannot be established on the basis of the consequences of that act and its alternatives alone, even in cases in which those consequences are perfectly predictable. Such theorists may also *believe* that the reasoning that goes into a choice is at least very important in determining its permissibility.

The argument can be made, however, that the Abortion Paradox *shows* that view to be correct. If, that is, we cannot find a plausible alternate solution to that paradox, then we will be forced to adopt an account of permissibility that is reasons-based at least in part. So a great deal is at stake here.

Still, even theorists who are confident that a reasons-based approach is correct may not be satisfied by McMahan's solution. They may not be convinced that the distinction in the lines of reasoned analysis triggered in the agent's mind by the distinction in orders of presentation in fact helps to explain the distinction in moral evaluation. But McMahan himself points out that it "almost impossible to believe" that the order of presentation can affect permissibility in so profound a way.[105] We can't, in other words, imagine why any plausible theory of permissibility would place any weight on *that* particular distinction.

3.2.4 As noted earlier, for purposes here I understand the Abortion Paradox to arise in the context of early abortion. It's in that context, I believe, that we find a puzzle that we badly need to unwind. For suppose that we are not content with McMahan's own solution to the Abortion Paradox and think that the scenarios

[105] McMahan (2006), p. 650. He makes this comment in connection with still another paradox he develops in the same paper: "If ... the woman can get an abortion today and take pill 2 tomorrow, she can permissibly avoid a future of chronic pain; but if her circumstances are such that she must take pill 2 today and cannot have an abortion until tomorrow, morality seems to require that she endure that future of chronic pain." Such results, McMahan says, are "almost impossible to believe."

really do have all their morally relevant features in common. And suppose that we at the same time accept the line of reasoning – which includes, I will argue, Pareto Plus – that takes us to the result that the early abortion is wrong in one of the two scenarios he describes. Then, we may be left to think that we have no choice but to conclude that early abortion is *wrong in the other scenario as well.* In contrast, if we found ourselves forced to accept the result that late abortion is wrong in both scenarios, we might not be so troubled. We might even find that a plausible result.

Moreover, as we shall see, it is implausible that the line of reasoning that takes us to the result that abortion is wrong could be limited to *late* abortion. If that line of reasoning really does show that *late* abortion is wrong in some cases, then it would seem to show that *early* abortion is wrong as well in exactly those same sorts of cases. Pareto Plus, to repeat, does give rise to a very tidy argument against early abortion. And that is so, even if McMahan's own argument is meant to apply just to the case of late abortion.

3.2.5 I thus propose a different way of dealing with the Abortion Paradox. I propose that we reject it. In particular, I propose that we reject the line of reasoning that takes us to the result that the early abortion is wrong.

But on what basis? I argue that at a critical juncture the Standard Pareto Principle is not on its own enough to justify the next step of the argument. I argue that we instead need Pareto Plus – and further that Pareto Plus is itself a weak link.

The Standard Pareto Principle is very intuitive and stands up to rigorous testing. But it applies *only if* the choices to be compared are same-people choices. The problem is that it is hard to construe the choices that McMahan needs to compare as same-people choices – or even as same-individual choices. The *early fetus* exists under both the early abortion choice and the fetal injury choice. The *pregnant woman* exists under both choices. And so do *many other individuals having moral status*. The problem is that *the child* does not. The child obviously exists under the fetal injury choice. But there is no basis for saying that *the child* exists under the early abortion choice as well. Indeed, the very point of the early abortion is to keep *the child* – that *person* – from ever coming into existence to begin with.[106]

We are thus left with Pareto Plus. Pareto Plus is clearly applicable in the scenarios McMahan describes. And we can't get around the fact that Pareto Plus on the face of things may well seem compelling.[107] For that reason, the Abortion Paradox – like the Mere Addition Paradox that precedes it and that it so closely resembles – is *genuinely* perplexing.

But still there is a problem. Pareto Plus is *far too questionable a principle* ever to even begin to suggest that we face paradox. Pareto Plus may well seem compelling on the face of things. Beneath the surface, however, it is not intuitive at all and does

---

[106]McMahan himself may think that the early fetus – or at least the fetus that he considers the subject of the Abortion Paradox – is identical to the child. I think it simplifies and clarifies things to think of the child as an additional person. Either way, a different metaphysics is not going to make the substantive moral debate go away. See note 143 below.

[107]Parfit (1987), pp. 419–441.

not stand up to any rigorous testing. And I will argue that a comparison between Pareto Plus and Variabilism makes it very clear just why that is so.

3.3 The Concern with Pareto Plus

3.2.1 As we shall see, it is clear that McMahan invokes a Pareto principle in his construction of the Abortion Paradox. A rough statement is this:

> *Rough Pareto Principle*. The fetal injury choice is *morally better* than the early abortion choice, *if* the fetal injury choice is better for a person, or other individual who has moral status (e.g. the early fetus), than the early abortion choice is *and* the fetal injury choice is not worse for any person, or any other individual having moral status, than the early abortion choice is.

This principle can be constructed in at least two ways. We can think of it as taking for granted that the fetal injury choice and the early abortion choice are same-people choices. Or we can think of it as applying even in the case where an additional person exists under the fetal injury choice and under that choice has an existence worth having. We can, that is, construct it as the Standard Pareto Principle. Or we can construct it as Pareto Plus.

But however we construct the Rough Pareto Principle, the Abortion Paradox itself arises as a function of the fact that we can, according to McMahan, *only intermittently rely on that principle to generate accurate permissibility results*. In other words, the fact that we learn that one choice is *morally better* than another sometimes, according to McMahan, means that the other choice is wrong and sometimes doesn't.

It seems, then, that McMahan would reject what we might call the *Simple Permissibility Principle*.

> *Simple Permissibility Principle*. If choice *a′* is morally better than choice *a*, then *a* is impermissible.

And we can speculate that he instead accepts a certain *schematic* for permissibility, one that when filled out will reflect, not just moral betterness, but other factors as well. We can call it the *Multifactorial Permissibility Principle*:

> *Multifactorial Permissibility Principle*. If choice a′ is morally better than choice *a*, then *a* is impermissible *unless* ___________.

This schematic contemplates that *a* will in some cases be permissible even though a′ is *morally better* than *a*.

3.2.2 The Rough Pareto Principle may seem plausible enough. After all, it is highly restricted. The fetal injury choice must be better for some person – or, as McMahan himself might put it, some *individual* having moral status. And the abortion choice cannot make things worse for *even one person*, or individual, than the fetal injury choice does.

It might then seem that the most contentious feature of the Rough Pareto Principle concerns its parenthetical assumption that the *early fetus* is the kind of individual who has at least some moral status. It is that assumption that means that the Rough Pareto Principle itself endorses the idea that the fetal injury choice being *better for the early fetus* than the abortion choice has *any moral significance at all.*

And certainly that is just an assumption. It is far from obvious that the fact that the fetal injury choice is *better for the early fetus* can make that choice *morally better* than the abortion choice. One fertilizer might be *better for the daffodils* than another without there being any suggestion that it's *morally better* to use the one fertilizer than the other.

So there is an issue here. McMahan assumes (for purposes of the argument) that the fetus (unlike the daffodil) has at least *some* "moral status" in its own right and that it is in the fetus's interest (though not in the fetus's *conscious* interest) to "continue to live" and eventually develop into a full-fledged person.[108] He thus assumes that, other things being equal, betterness-for-the-early-fetus counts *in favor of* the fetal injury choice; other things begin equal, it makes the fetal injury choice *morally better than* the choice of early abortion.

But this issue is effectively mooted in the broader discussion. That is so, for three reasons. First, McMahan's own tactic is to grant that the early fetus has *some* moral status without accepting that the early fetus has *anything like* the moral status *we ourselves* have. After all, the early fetus is not yet sentient. It has never engaged in that *first thinking experience* that – in my own view – seems to signal the coming into existence of an individual who, if not technically yet a *person* according to McMahan, has a moral status that comes close to our own. Thus, when push comes to shove – when we must choose between protecting the interests of the pregnant woman and protecting the interests of the early fetus – McMahan can easily take the position *that it is the pregnant woman who must prevail.*

In the end, we will want to come back to this issue. My own position is that it is the pregnant woman who must prevail even when push does not come to shove. Even, that is, when she has no interest at all at stake in how the choice of early abortion is made, in any typical case the early abortion is permissible for her. She can permissibly choose early abortion, in any typical case, for a good reason, a bad reason or no reason at all. But that is an issue for Chapter 5.

Second, even if we grant the assumption that the early fetus has some moral status, we are still going to find that the Rough Pareto Principle – *depending on how it is constructed* – is *engulfed in a controversy of mammoth proportions*. We don't, in other words, *need* to reject the assumption that the early fetus has some moral status in order to reject the Abortion Paradox.

And third: Our rejecting the assumption that the early fetus has some moral status does not offer an alternate basis for rejecting the Abortion Paradox. That is,

---

[108] McMahan (2006), p. 627. He is even willing to say that the fetus has an "interest" – though obviously not a *conscious* interest – in the agent's choosing fetal injury over abortion. McMahan (2006), pp. 628 and 636–658.

the Abortion Paradox, with only slight modifications in Pareto Plus, will still go through, even in the case where we jettison the assumption altogether.[109]

3.2.3 The Rough Pareto Principle thus seems plausible – and its parenthetical assumption something we will not, for purposes here, contest. McMahan then uses that principle – in combination with the Multifactorial Permissibility Principle, or so I am speculating here – to construct the Abortion Paradox.

But as noted earlier the Rough Pareto Principle is itself subject to two constructions. There is the very plausible Standard Pareto Principle. And then there is *Pareto Plus*. I will argue that Pareto Plus is false.

A door is thus open to a critique of the Abortion Paradox itself. If that critique succeeds, then we will have no need to try to discern moral import in seemingly unimportant factual distinctions – in particular, in distinctions in the order of presentation of options to the agent – between the scenarios. Instead, we can reject the paradox.

3.2.4 The position I take against Pareto Plus is itself contentious (at least among consequentialists, of which I am one). And I want to explain very briefly why I think it's worth taking.

Many theorists – perhaps McMahan himself, and certainly, more recently, Michael Huemer – will consider Pareto Plus every bit as compelling as the Standard Pareto Principle.[110] By hypothesis, the child's life is worth living under the fetal injury choice, which means that there is no room to argue that the fetal injury choice makes things worse for the child than abortion does. If anything, the fetal injury choice would seem to make things *better* for the child than abortion does. Moreover, also by hypothesis, the fetal injury choice is not worse for *the pregnant woman herself* or for *anyone else* than the abortion choice. Finally, the fetal injury

[109] The reason that the Abortion Paradox, with only slight modifications in Pareto Plus, will go through even in the case where we jettison the assumption altogether is that all we really need to get to paradox is the result that the fetal injury choice is equally as good as the early abortion choice. Since we know that the fetal injury choice is itself wrong, if the two are "equally good," then that should tell us that the early abortion choice is wrong as well. McMahan can then develop the scenarios just as he does under the original Abortion Paradox, and we will again find that the early abortion is permissible in one of the scenarios and wrong in the other.

[110] For further discussion of principles closely related to what I am here calling Pareto Plus, see Torbjörn Tännsjö (2004), pp. 219–237, and Michael Huemer (2008), pp. 899–933. Huemer even goes so far as to describe the nonstandard principle as an "ethical axiom" – and he uses it as part of a "proof" to the repugnant conclusion. Thus he accepts what he calls the *Benign Addition Principle*: "If worlds *x* and *y* are so related that *x* would be the result of increasing the well-being of everyone in *y* by some amount and adding some new people with worthwhile lives, then *x* is better than *y* with respect to utility." Huemer (2008), pp. 901–902. He says this principle is supported by what he calls the *Modal Pareto Principle:* "For any possible worlds *x* and *y*, if, from the standpoint of self-interest, *x* would rationally be preferred to *y* by every being who would exist in either *x* or *y*, then *x* is better than *y* with respect to utility." Huemer (2008), p. 903. Perhaps there is a subtlety relating to the phrase "with respect to utility" that I am overlooking; perhaps that phrase means that his entire argument in favor of the Repugnant Conclusion is really just the argument that the choice that leads to more utility in the aggregate is better "with respect to utility" in the sense that it leads to more utility in the aggregate. But that, of course, would be to trivialize his results. But if he means anything more than that, then there is a serious question whether the principles he puts to work are truly axioms or simply falsehoods.

choice is *better for the early fetus* than the abortion choice. It may thus seem a simple matter to conclude – exactly as Pareto Plus suggests – that the fetal injury choice is *morally better* than the abortion choice.

In what follows, however, I will argue that, appearances notwithstanding, Pareto Plus fails to take *fully* into account *all* the relevant moral interests. We *think* we are fully taking into account all the interests that can potentially have any moral significance at all in determining the moral permissibility of a certain choice. In fact, however, the very structure of Pareto Plus, I will argue, *insures* that we have screened out at least some of those same interests.

Variabilism underlines this point. The very structure of Pareto Plus insures that we have screened out some of the interests of *the possible future child* – and, in particular, the interest that child has in coming into a better rather than a lesser existence; the interest that child has in the agent creating more wellbeing for that child rather than less. That is an interest Variabilism can clearly take into account. When we compare one choice against another, Variabilism underlines that losses that lie beyond just the outcomes of those two choices can have moral significance for purposes of determining the permissibility of the one choice. A world w' may be better for a person than a world w, yet that person may still incur a loss at w' – relative to some third world w'' – that Variabilism will recognize as morally significant. Any otherwise plausible permissibility theory will take that loss into account in determining permissibility. But Pareto Plus does not. Plausibly, we should reject Pareto Plus.

3.4 The Abortion Paradox

3.4.1 We now turn to the paradox as McMahan himself lays it out. He compares two scenarios:

Scenario I (the *one-form* scenario). A pregnant woman, if not treated, will experience permanent *mild* pain. Her doctor gives her three treatment options – *Pill 1*, *Pill 2* and *No Pill*. She is told to make her choice by checking off one of the three options on a *single official form*. The treatment options have the following effects (and no others):

Pill 1: cures the woman; *injures* and then *painlessly kills* the early fetus.

Pill 2: cures the woman; *injures* but does *not kill* the early fetus; the child the fetus develops into suffers permanent *moderate* pain.[111]

No Pill: the woman, uncured, suffers permanent *mild* pain; the fetus is *not injured*; the child the fetus develops into is healthy.

Scenario II (the *two-form* scenario). Scenario II is just like Scenario I except that the woman is told to make her choice by completing (up to) two official forms. On the first form, she checks off *Pill* or *No Pill*. If she chooses *Pill*, then, on the second form, she checks off *Pill 1* or *Pill 2*.

[111] McMahan (2006), p. 627.

The wellbeing distributions – facts that we think are clearly relevant to the moral analysis of the two scenarios – for both Scenario I and Scenario II are the same:

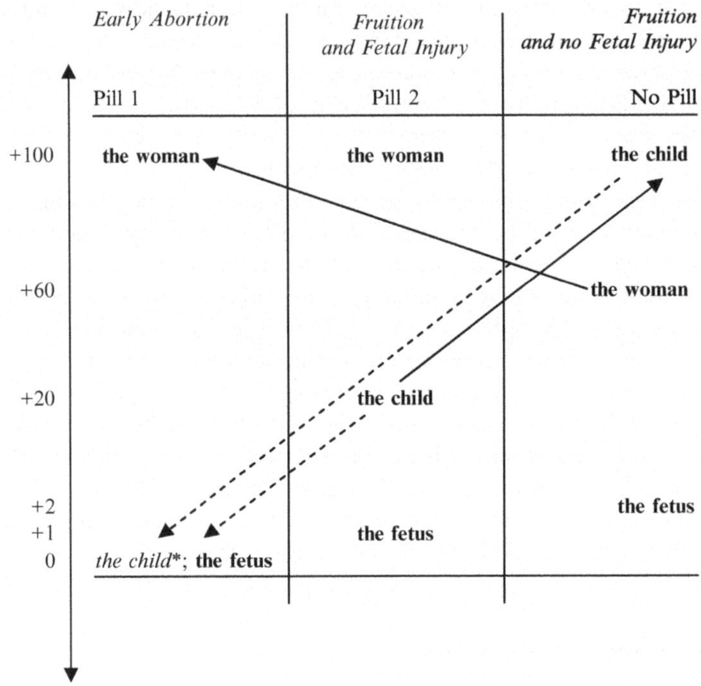

Case 3.1 The Abortion Paradox

This graph shows the different wellbeing levels the parties are accorded under the different choices. It also sums up just which of McMahan's underlying assumptions about the case I am granting for purposes here and which I am not. McMahan accepts that the fetus has some slight interest in "continuing to live" – an interest that is in fact satisfied in the case where the fetus *does* continue to live. That idea is represented in the low but positive level of wellbeing the fetus is assigned under Pill 2 and the still low but very slightly higher wellbeing level the fetus is assigned under No Pill. Pill 2 is, in other words, better for the fetus than Pill 1 is, and so is No Pill. McMahan also accepts that the fetus has *some* moral status – though not as much as you or I. Thus, the fact that Pill 2 is better for the fetus and worse for no one at all than Pill 1 is means, according to McMahan, that Pill 2 is *morally better* than Pill 1. The assumption that I am *not* making for purposes here is that *the early fetus is identical to the child*: the child, instead, is an *additional person*, a person who exists under Pill 2 but not under Pill 1.[112]

McMahan's moral evaluation, then, of Scenario I – the one-form scenario – consists of claims (1)–(3).

[112] McMahan may contest this point. But see note 143 below.

1. The choice of Pill 1 – the early abortion choice – in Scenario I is permissible.

Claim (1), as McMahan suggests, seems plausible. We may well think that the woman need not sacrifice her own interests – by choosing No Pill – for the sake of the *early fetus*. McMahan assumes that the early fetus, though not a person, has some "moral status" and a weak "interest" (though no conscious interest) in "continuing to live."[113] But he also thinks that its moral status is "lower" than the woman's.[114] Moreover, "its death would not be a terrible misfortunate for it."[115] Nor does it seem plausible that the woman must sacrifice her own interests for the sake of her *possible future child*. We have no moral reason, McMahan writes, to "create new interests" that we then must satisfy.[116] In contrast, the woman's interest in avoiding even mild permanent pain is strong. Moreover, she has full moral status. The upshot is that the woman "is not required to accept the chronic pain."[117] She is not required, that is, to choose No Pill. Pill 1 – early abortion – is, accordingly, permissible.

It is clear that, for McMahan, the comparison between Pill 1 and No Pill is at least part of what justifies claim (1). "First we compare taking Pill 1 with taking No Pill."[118] Balancing the relevant interests, McMahan concludes that "Pill 1 is permissible."[119]

McMahan's argument at this juncture may seem truncated. For we know we can't infer permissibility *on the basis of a single pair-wise comparison* if still other options remain available to the agent. Suppose that Pill 1 *is* morally better than No Pill in virtue of what it does for the woman and notwithstanding what it does to the fetus. If a third option is *still better* – for, say, both woman *and* fetus – we are surely forced to say that Pill 1 is wrong.

It is possible, however, that McMahan is counting on us to accept that a little truncation at this point is unobjectionable. And that seems right, since any protracted inspection of the only remaining option – Pill 2; the fetal injury choice – is surely pointless in view of what Pill 2 does to *the future child*. We may thus be very pleased to stipulate that Pill 2 is clearly wrong – that it is "off the table" and that the only arguably permissible choices are Pill 1 and No Pill. But we know that there must be at least one permissible alternative. Pill 1 being deemed morally better than No Pill, it seems safe to say that Pill 1 is permissible.

For roughly these very reasons, claim (2) seems clearly acceptable.

---

[113] McMahan (2006), pp. 636–639.

[114] McMahan (2006), pp. 636–639. Going forward, I frequently reference McMahan's discussion of the three option case. McMahan (2006), p. 651.

[115] The fetus's interest in "its own future life" is weak – in virtue of the fact that it is "only distantly psychologically related to the person it might become." McMahan (2006), p. 628. Later on, the fetus – and the subsequent person, that is, the child, whom the fetus is, in some sense, "identical" to and develops into – will have an increasingly strong interest in its "future life." McMahan (2006), p. 627.

[116] McMahan (2006), p. 631.

[117] McMahan (2006), p. 636.

[118] McMahan (2006), p. 639 (capitalization added here and throughout).

[119] McMahan (2006), p. 651 (referring back to the Pareto Paradox and the three options case, discussed at pp. 636–639).

2. The choice of Pill 2 – the fetal injury choice – in Scenario I is impermissible.

Laying the argument out in more detail, McMahan, again, balances the relevant interests. But here the interests of *the future child* are front and center. Thus: "the pain [the woman] would … cause her child to suffer [by taking Pill 2] would be more severe and of longer duration than the pain she would avoid. It is … impermissible for her to avoid a lesser harm to herself by inflicting a greater harm on her own child…."[120]

The comparison, in other words, between No Pill and Pill 2 tells us that Pill 2 is wrong: No Pill shows that the agent has a (far) better way of accommodating the various interests than what we see under Pill 2.

Finally:

3. The choice of No Pill in Scenario I is permissible.

No Pill leaves the fetus's interests in continuing to live *and* the child's interest in a healthy existence entirely protected – and, by hypothesis, makes things worse for no one *other than the woman herself.* "It is obviously permissible for her to take neither pill" – despite the fact that that choice leaves the woman in permanent mild pain. This conclusion can be questioned, but for the moment we let it go.[121]

Claims (1)–(3) comport fairly well with what we intuitively want to say about this case. The woman is permitted to have the early abortion – at least in the case where carrying the fetus to term will cause her to suffer permanent mild pain and, very plausibly, in many other cases as well. But she is *not* permitted to *injure* the fetus in a way that allows it to survive in the case where the child the fetus would develop into will then suffer permanent moderate pain.

But things change radically, according to McMahan, when we turn to Scenario II – the two-form scenario. Scenario II begins with the woman's perusal of a form that gives her the choice between Pill and No Pill. She may choose either to take a pill that would treat her permanent mild pain or not to take a pill and, accordingly, to suffer that pain. Understanding the effects of both pills and "believing that Pill 1 is permissible," she chooses Pill.[122]

She then turns to a second form, where she is now required to say which pill she will take. And it is here, according to McMahan, that we obtain the unexpected claim (1′):

1′. The choice of Pill 1 – the early abortion choice – in Scenario II is impermissible.

---

[120] McMahan (2006), p. 633. See p. 651 (referring to earlier discussion at pp. 633–636).

[121] McMahan (2006), p. 636. There is much more to be said about claim (3), as well as claim (3′) below. At least, the argument for those claims needs to be reworked to accommodate the case where the agent is not the pregnant woman herself at all but rather another party, say, a spouse or a physician. But there is a deeper point in this form of reverse partiality – where the agent need *not* take into account his or her own interests in determining what ought to be done. It will mean that, on the same facts and calculating by reference to the same moral theory but relativizing to agents, we generate inconsistent results. The doctor, for example, will, or should, calculate that No Pill is clearly wrong, given the option of Pill 1 and given what No Pill does to the woman. In contrast, the woman, McMahan suggests, can deny her own interest and reach the result that No Pill is perfectly permissible. But which assessment is right? The better approach is for each agent to calculate taking into account *all* morally significant interests. See note 204 below.

[122] McMahan (2006), p. 651.

But how can Pill 1 lose permissibility between Scenario I and Scenario II? How can claim (1) and claim (1′) both be true?

McMahan argues as follows. Once the woman has completed the first form and has chosen (permissibly, she, and we, might have thought) Pill, her interests are "not at stake: whichever choice she makes, her condition will be cured. So she ought to be guided by what would be *best for the fetus*, which … is Pill 2."[123] "Pill 2 is, in short, *Pareto superior* to Pill 1."[124] Therefore, Pill 1 is impermissible.

Then, for reasons already noted, (2′) seems correct:

2′. The choice of Pill 2 – the fetal injury choice – in Scenario II is impermissible.[125]

Moreover:

3′. The choice of No Pill is permissible in Scenario II.[126]

We can now simply point out that, if (1′)–(3′) are true, so is the idea that the woman's "*only* permissible option" in Scenario II is No Pill[127]:

4′. The choice of No Pill is obligatory in Scenario II.

Since (4′) follows, given (1′)–(3′), from standard deontic axioms, the only real surprise here is (1′).

3.4.2 It was noted when we first turned to Scenario II that, when the woman peruses the initial form, which gave her (just) the choice between Pill and No Pill, she chooses Pill, thinking it permissible since she realizes that she will be free to choose Pill 1 and avoid the odious Pill 2. One might have thought that fact would complicate the foregoing analysis in some way – that the fact that she's *now made the mistake* would open the door to a re-evaluation of claim (1′). But it doesn't, according to McMahan. (1′) is meant to hold even *after* the woman has completed the first form and chosen Pill rather than No Pill. There is no suggestion here that the woman, once she turns to the second form, has *no* morally permissible options. The thought rather is that her choice on the first form was *revocable*. Once, that is, she peruses the second form and comes to understand that Pill 1 and Pill 2 *are both wrong*, she can change her choice to No Pill.

This back-stepping is what morality will force the woman who isn't "particularly clever" to do: if she mistakenly thought, as she completed the first form, that the choice of Pill *was* permissible – she "ought now to write back, revoking the choice she made" before.[128] If she *is* particularly clever, she'll presumably do things right the first time around and choose No Pill starting out.

---

[123] McMahan (2006), p. 651 (emphasis added).

[124] McMahan (2006), pp. 639 and 641 (emphasis added).

[125] Again: "She ought not to avoid [mild] chronic pain at the cost of causing her own child to suffer a worse form of chronic pain." McMahan (2006), p. 651.

[126] The idea that No Pill is permissible is especially troubling in Scenario II; it seemed more or less plausible in Scenario I in virtue of the fact that in that case the woman was also said to have Pill 1 as a permissible option. See note 204 below.

[127] McMahan (2006), p. 651 (emphasis added).

[128] McMahan (2006), p. 651.

Either way, McMahan stands behind (1′). The upshot is that all three options all remain *available* to the agent and *known* to the agent at *each discrete moment* throughout both Scenario I and Scenario II.

3.4.3 And so we have a paradox. From Scenario I to Scenario II, the fetus is no closer to consciousness; the pain the woman will suffer without any pill remains both permanent and mild; the pain the child faces remains both permanent and moderate if Pill 2 is chosen; and no one else has any interest in the matter whatsoever. How, then, could it possibly be permissible for the woman to take Pill 1 in Scenario I but not in Scenario II? How could the early abortion choice possibly be wrong in Scenario II but not in Scenario I?

3.5 McMahan's Solution: Order of Presentation

3.5.1 To resolve the Abortion Paradox, McMahan points to the distinct "orders of presentation" of the relevant options. Whether the options are presented to the woman concurrently, as in Scenario I, or sequentially, as in Scenario II, "affects what the … relevant considerations are at the time of choice. It therefore affects what *reasons* one has for or against acting in certain ways …," which in turn affects permissibility.[129] "[B]y first comparing Pill 1 with No Pill [as in Scenario I] and then comparing Pill 2 with No Pill, we reach the conclusion that [in Scenario I] Pill 1 is permissible but Pill 2 is not. But we have not compared Pill 1 directly with Pill 2 [as we do in Scenario II]. When we compare Pill 1 with Pill 2, we find that Pill 2 is better for the fetus [than Pill 1] and not worse for anyone else."[130]

As the order of presentation shifts, thus so does the agent's moral analysis – the reasons the agent has for acting one way rather than another. And as the agent's moral analysis shifts, so does the permissibility of the choice.

3.5.2 Two points should make us wonder whether the best way out of the Abortion Paradox really has anything to do with the distinction in order of presentation. First, it doesn't help us avoid inconsistency on any sustained basis. After all, if the woman in Scenario II initially chooses Pill rather than No Pill, and then realizes

[129] McMahan (2006), p. 652 (emphasis added). The fuller quote is as follows: "an alteration in the order of choice … affects what *the options and* relevant considerations are at the time of choice. It therefore affects what *reasons* one has for or against acting in certain ways …." I have omitted the italicized portion, since it is clear that that language cannot be intended to be read literally and is meant instead to refer to the particular options the agent happens to be comparing (on a pairwise basis) at a given time. After all, if *the sets of options* available to the agent (the *feasible sets*) at the critical time just before choice were truly distinct – if, for example, in Scenario I the woman's options were the three we have listed, but in Scenario II, the woman's options were limited to just Pill 1 and Pill 2, we could at least understand (though we might well dispute) why Pill 1 is permissible in Scenario I but not Scenario II. There would be no contradiction. It is, moreover, clear in McMahan's discussion of the twin scenarios that all three options are available to the agent in each.

[130] McMahan, p. 640. This comment is made in connection with his discussion of what he calls the *Pareto Paradox*. The Abortion Paradox builds on that discussion.

that both the choices she faces are impermissible – Pill 2 because of what it will do to the early fetus, and Pill 1 because Pill 2 is Pareto superior to Pill 1 – she "ought ... to write back, revoking the choice she made on the first form."[131] That means that at the time the woman begins to complete the second form, all three options are available to her. We then seem to face the same old inconsistency – and now have no distinction in order of presentation to explain it away.

Second, the idea that the line of reasoning triggered within the agent's own mind can help to determine the permissibility of a given choice is plausible *only if* it is understood that that line of reasoning itself conforms to certain minimum standards. Where there are inconsistencies in the analysis, the critical distinctions are themselves arbitrary or the analysis itself is badly incomplete, the permissibility results that the distinct lines of reasoning take us to cannot be relied on. But McMahan's proposal asks us to rely on just such results. That proposal explains, thought by thought, how the agent reaches one result in one case and another result in the other case. Yet it leaves us without an *understanding* of why the Paretian account that tells us that Pill 2 is morally better than Pill 1 is given pride of place in Scenario II but not attended to at all in Scenario I.

It is unclear, in other words, why we should not consider Pill 2 morally better than Pill 1 not just in Scenario II but in Scenario I as well. McMahan himself suggests that the fact that one option is *morally better* than another does not always mean that that other option is *impermissible*. A filled-out *Multifactorial* Permissibility Principle will, in other words, take into account both moral betterness and facts regarding the order of presentation. But we may be left puzzled why that is so – *why* the distinction in order of presentation carries so much weight.

It is similarly unclear why, still elsewhere in McMahan's paper, the analysis we see in Scenario II is considered objectionable in its own right. Why, in other words, should we think that Pill 1 is morally worse than Pill 2, in a case where it is already established that Pill 1 is permissible and Pill 2 is not. McMahan calls this the *Pareto Paradox*[132]; and he says that "[t]his is a conclusion that defenders of abortion rights will want to reject."[133] But on what grounds?

The point can be put the other way around. McMahan's proposal leaves us without an account of why the leap of deontic logic we see in Scenario I – the leap that ignores the option of Pill 2 and allows the agent to conclude, on the basis of the single pair-wise comparison between Pill 1 and No Pill, that Pill 1 is permissible – can be made there but is barred in Scenario II. Why, that is, does our recognition of the obvious wrongness of Pill 2 allow us to truncate the argument in Scenario I – and conclude immediately, given that Pill 1 is morally better than No Pill and given that there are no other *permissible* options, that Pill 1 is permissible – but not in Scenario II? Why does the fact that Pill 2 is better than Pill 1 mean that we can truncate the argument in Scenario I but not in Scenario II?

---

[131] McMahan (2006), p. 651. If one option is revocation, it is unclear why her interests are *not* at stake at the second form stage of Scenario II.

[132] McMahan (2006), pp. 639–641.

[133] McMahan (2006), p. 641.

3.5.3 McMahan's solution could be viewed as providing us with a *psychological* explanation why someone might come to the result that Pill 1 is permissible in Scenario I but wrong in Scenario II. But to resolve the Abortion Paradox along the lines that he is proposing, what we really need is the filled-out Multifactorial Permissibility Principle. We need a principle that we can test and that can explain to us at an intuitive level why the moral betterness of Pill 2 is material to the analysis when the three options – which are all *available* to the agent and *known* to the agent at *each discrete moment* in both Scenario I and Scenario II – are presented to the agent all at once but not when they are presented to the agent across time. That is a lot to ask for. But it would take us a long way toward accepting McMahan's own solution to the Abortion Paradox.

3.6 An Alternative Solution: Variabilism

3.6.1 McMahan's solution to the Abortion Paradox is based on the identification of a factual distinction between the two scenarios – a factual distinction that in the end does not seem to explain the distinct results. The solution I will examine in this part 3.6 *rejects the paradox* by rejecting the argument to claim (1′) – the argument, that is, that the early abortion choice is impermissible in Scenario II.[134]

My solution has two parts. The first part acknowledges that there is something about the woman's choice of Pill in Scenario II that does raise an alarm of a certain sort. I argue, however, that it is the kind of alarm we see whenever the options available to the agent are spread out across time. And I argue that the concern that is signalled is one that we know how to address.

The second part objects that the argument to claim (1′) itself fails. Here, McMahan's Paretian analysis is at stake. I argue that it is Pareto Plus, not the Standard Pareto Principle, that is invoked in McMahan's analysis, and that that principle, in coming to the result that Pill 2 is morally better than Pill 1, fails to take into account *what Pill 2 does to the future child.* Pill 2 clearly harms that child; it imposes on that child a morally significant loss; it thwarts an important interest that that child has. Pareto Plus, however, effectively *screens all of that out.* It assesses Pill 2 as morally better than Pill 1 without ever coming to grips with *just how bad Pill 2 is.* Once we bring the child back into the equation, we can, I shall argue, easily see the flaw in Pareto Plus and avoid the result that Pill 2 is morally better than Pill 1. At that point, we will be left under no pressure at all to conclude that Pill 1 is impermissible, in either Scenario I or Scenario II.

3.6.2 When choices are scattered across time, as they are in Scenario II, we might feel that we have lost a certain guaranty of moral rectitude that we have when

[134]Another basis for rejecting the Abortion Paradox would be to reject claim (1) – to say, that is, that Pill 1 is impermissible in Scenario I. On that way of dealing with the paradox, we would accept McMahan's Paretian analysis across the board, rather than simply intermittently, as McMahan himself does. The problem with that solution to the paradox is – as noted earlier, and as we explore in more detail below – that that particular Paretian analysis is seriously flawed.

they are not, as in Scenario I. We might thus feel that without that guaranty it's important to retain claim (1′).

More specifically, in Scenario II, we could more comfortably reject (1′) if we could be *sure* that the woman will *not* choose Pill at the earlier time (when presented with the first form) and then at the later time (when presented with the second form) choose Pill 2, which we think is clearly wrong. If she's going to do *that*, then, we think, she ought not have chosen Pill to begin with. And one way of guarantying that she not do that is to make the choice of Pill impermissible from the start.

It may seem that the very guaranty we are looking for is built into Scenario I. There, the woman's only vehicle for choosing Pill is by committing herself to Pill 1 or committing herself to Pill 2. At that point, Pill 2 *really is* an option for her. However, by dint of moral analysis we agree that she is *obligated* to avoid Pill 2. And that's our guaranty: that, though Pill 2 is available to the woman as an option, moral law *obligates* her not to take it.

But it would be a mistake to think that Scenario II does not incorporate exactly that guaranty. Without changing the case, we can't rewrite Scenario II in such a way that she is *physically* restrained from choosing Pill 2. But nor is that how Scenario I is written. What we can do, however, is rewrite Scenario II as a standard problem of conditional obligation. Once we do that, we can make it clear that the woman is *obligated* to *avoid* Pill 2 in Scenario II – exactly as she is in Scenario I.[135]

Thus we can say: the woman's choice of Pill at the earlier time (t1) in Scenario I is *impermissible* at any world at which the woman chooses Pill 2 at the later time (t2), and *permissible* at any such world at which she chooses Pill 1 at t2. Suppose, then, that at a particular world – say, the *actual* world – the woman chooses Pill 2 at t2. Her choice of Pill at t1 at that world is impermissible. That is so, in virtue of the fact that, at t1 and at that world, the option of taking No Pill at all exists as an alternative for the woman. It is the availability of that option at that time and that world that makes her choice of Pill wrong. It is, in other words, the availability of that option at that time and that world that shows that the injury her future child will eventually suffer is both perfectly avoidable and unjustifiable.[136]

It's true that, on this way of looking at things, the relevant condition on the permissibility of Pill may not be *known* to be satisfied or failed – again, at the actual world – until *after* that choice has been performed. More precisely: it won't be settled whether that condition is satisfied or failed until it is settled whether the actual world includes the agent's choosing Pill 1 at t2 or Pill 2 at t2 – until it is settled, in other words, just how the future will actually unfold. No matter. We still obtain the result that we think is correct: the choice of Pill at t1 is wrong at any world at which the woman then proceeds then to choose Pill 2 over Pill 1 at t2.

---

[135] The way of looking at the two scenarios that I describe in this part 3.6.2 is based on Fred Feldman's work on conditional obligation. See Feldman (1986), pp. 86–95 and (1990), pp. 309–341.

[136] Moreover, this analysis holds whether the woman's choices at t2 are exhausted by Pill 1 and Pill 2 or not – whether, that is, she has the option of revoking her choice at t2 or not. We simply say that her choice of treatment at t1 or at t2 is impermissible if she chooses Pill 2 at t2.

On inspection, we can see that Scenario I is also a problem of conditional obligation just as Scenario II is. In fact, what we should say in both Scenario I and Scenario II is that the choice of Pill over No Pill is permissible *but* that it is not *categorically* or *unconditionally* permissible in either case. Rather, it is *conditionally* permissible. As Fred Feldman suggests, "A number of philosophers have suggested that statements of conditional obligation are … tied down to the passage of time. … I think this suggestion is erroneous."[137] We should instead say that, for Scenario I, the choice of Pill at t1 is permissible *if* Pill 1 and not Pill 2 is chosen *at t1*. For Scenario II, the choice of Pill at t1 is permissible *if* Pill 1 and not Pill 2 is chosen *at t2*.[138]

3.6.3 If our sole concern were to insure that we do not generate the result in Scenario II – or Scenario I – that the choice of Pill is permissible in cases where the woman then chooses Pill 2, we would be done. But we have another concern as well. And that is that the Paretian analysis that McMahan uses to justify claim (1′) is on the face of things compelling. To truly resolve the puzzle, we need to show just how it goes awry.

As between Pill 1 and Pill 2, the woman's interest in avoiding permanent mild pain is not at stake. Her interests are satisfied so long as she takes one pill or the other. Pill 2 thus is not worse for the woman than Pill 1 is. Nor is anyone else made worse off by Pill 2 than by Pill 1. That includes the early fetus. But it also includes the *possible future child*. After all, by hypothesis, the case is not one in which the fetal injury is so severe that the child would have an existence *less* than worth having; this is not a case involving a genuinely wrongful life. And certainly throughout this book I have accepted that – and in fact believe that – leaving a person out of an existence worth having causes that person to incur a *loss*. It creates less wellbeing for that person when agents could have created more; it makes things worse for that person when agents could have made them better. And I am certainly not going to jettison that view now. If the fetal injury choice is morally worse than the early abortion choice, it is not going to be worse in virtue of the fact that we somehow (suddenly) can't compare what never existing at all has to offer the child and what existence has to offer the child. We can still make that comparison: the child has zero wellbeing at the world where that child never exists at all, and some positive level of wellbeing at the world where the child has an existence worth having.

Something like the Rough Pareto Principle, on these facts, seems easily to imply that Pill 2 is *morally better* than Pill 1 is.

McMahan's argument proceeds on that basis. At least as between just Pill 1 and Pill 2, he writes, the woman "ought to be guided by what would be best for the fetus, which … is Pill 2."[139] Given, however, that the woman "ought" choose Pill 2, it is then impermissible for her to choose Pill 1.

3.6.4 This appears to be a tidy argument against Pill 1. Beneath the surface lurk, however, some deep and unsettled matters. Just how deep and unsettled those matters

---

[137] Feldman (1986), p. 88.

[138] We thus avoid the difficulties inherent in assessing "plans or policies" in place of the acts themselves. See McMahan (2006), p. 644.

[139] McMahan (2006), p. 651.

are is suggested by certain analogies between the Abortion Paradox and still another collection of claims in respect of which we can be very sure something has gone awry – that is, the *Mere Addition Paradox*.[140]

Where boldface indicates that the person *p* exists under the specified act and that the person *q* exists under acts A+ and B, and the italics coupled with the asterisk that *q* never exists at all under A, the Mere Addition Paradox can be graphed as follows:

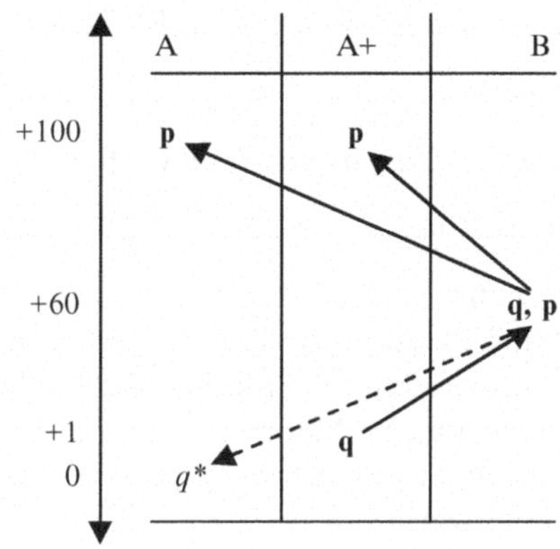

Case 3.2 Mere Addition Paradox

The paradox arises when we ask, about the otherwise unobjectionable A, how the "mere addition" to A of a life that is worth living, an addition that yields the further option A+, can make things "*worse*."[141] We seem compelled to say that it can't – and that, instead:

1″. A+ is at least as good as A is.

Now, in the Abortion Paradox, we face the stronger claim that the choice of fetal injury is morally better than the choice of early abortion. But that's not necessarily because we are there using a stronger version of Pareto Plus.[142] The difference is, instead, the early fetus: that individual is assumed to have some moral status, and fetal injury is better for that individual than early abortion is.

[140] Now, the analogies hold only if we consider the child itself an *additional person*. That is not, as we've noted earlier, McMahan's view. According to him, the fetus and the child are one and the same. Consequently, since the fetus exists under each option, so does the child. But see note 143 below.

[141] Parfit (1987), p. 430.

[142] Dasgupta discusses the stronger version of Pareto Plus. On that version, the mere addition of the person whose life is worth living makes the outcome *better*. The Abortion Paradox and the Mere Addition Paradox only require weaker versions of the principle. We can, that is, consistent with both those paradoxes consider the addition of the child to constitute a neutral change – to make things neither better nor worse. What decides, in the Abortion Paradox, that things are better is that they are *better for the early fetus*. What decides, in the Mere Addition Paradox, that the addition does not make things worse, is that things are not made worse for anyone at all.

But we then compare B against A+. And we seem compelled as well – whatever our normative instincts – to say the following:

2″. B is better than A+ is.

Finally, unless we are unreformed Totalists and consider betterness between acts nothing more than whether one act yields more wellbeing on an *aggregate* basis than another, we may well also have the intuition that we are not obligated to bring additional persons into existence *for their own sake*. That is:

3″. A is *at least* as good as B.

Assuming the transitivity of *at least as good as*, we then infer:

4″. A+ is at least as good as B is

But (4″), given just asymmetry, contradicts (2″).

Now, the facts underlying the Abortion Paradox are more complex. McMahan's moral analysis must take into account the plights of both the woman *and* the early fetus. Still, from the accounts that McMahan provides of Scenarios I and II – taken in combination – we seem able to extract betterness results that parallel the Mere Addition Paradox in striking ways:

1*. Pill 2 is at least as good as Pill 1 is.

(1*) derives from the Paretian analysis McMahan offers of Scenario II. His claim is that Pill 2 is morally *better* than Pill 1 in view of Pill 2′s effects on the fetus. Moreover, we agree that:

2*. No Pill is better than Pill 2 is.

We agree, that is, that it is better for the woman to suffer permanent mild pain than for the future child to suffer permanent moderate pain – and that Pill 2 is itself wrong.

From McMahan's argument for Pill 1 in Scenario I (obtained when the woman "first" compares Pill 1 against No Pill), we obtain:

3*. Pill 1 is at least as good as No Pill is.

Here we allow the woman's interest in avoiding permanent mild pain trump the early fetus's interest in continuing to live – and we set aside the possible future child's interest in coming into existence.

Transitivity, finally, yields (4*):

4*. Pill 2 is at least as good as No Pill is,

which, again by asymmetry, contradicts (2*).

As noted, I assume here that we *can* extract certain betterness results from McMahan's analysis. Moreover, the fact that there exist parallels between the Abortion Paradox and the Mere Addition Paradox does not, on its own, support the view that claim (1′) is false – or that McMahan's Paretian analysis is defective. At the same time, however, the analogies between the Abortion Paradox and the Mere Addition Paradox are hard to ignore. It should at least alert us to the fact that the argument that McMahan uses to obtain the result that Pill 2 is *morally better* than Pill 1 is – better, that is, than Pill 1 is, in a sense that itself has moral significance – and that Pill 1 is, accordingly, impermissible requires a *very* close look.

Both paradoxes underline the fact that lines of reasoning that can seem straightforward and unproblematic in ordinary contexts – *same-people* contexts – are *anything but* whenever additional persons and their plights enter the picture. We may *think* we are fully taking into account all the interests that can potentially have any moral significance at all in determining whether one choice is at least as good as the other – and such that if the other choice is permissible so is the one – when the very structure of our approach in fact *insures* that we have screened out at least some of those very same interests.

3.6.5 To lay the groundwork for an alternate solution to the Abortion Paradox, we really just need one claim: that, to compare two acts a and a' – whether A and A+ in the Mere Addition Paradox or Pill 1 and Pill 2 in the Abortion Paradox – in respect of their moral betterness, and to reach *any legitimate result* on the permissibility of either a or a', it is critical to take into account, not just how *each person* p who does or will exist under both a and a' is affected by a and a', but *also how each additional person* q – each person who exists under a' but not a – *is affected by any third alternative act* a''.

The intuition that grounds this claim is that we otherwise cannot fully assess *just how bad* a' is because we otherwise simply screen out *q's plight in* a'. Any such blind spot is especially dangerous when a' is *better for* some person p and worse for no person p who exists under both a and a'. In any such case, we will be left to think that a' is better than a and that, accordingly, a is wrong. Of course, the bare fact that q is better off under a'' than a' will not, *on its own*, show that a' is wrong or that a (or a'') is better than a'. But that q is better off under a'' than a' does seem a fact of high moral significance. That comparison *on behalf of* q does reveal something *against* a' that we would otherwise miss and that would otherwise leave us to focus exclusively on how a' (and a'') compares to *a on behalf of* p (who exists under both).

What it reveals is that q *incurs a loss* under a' (since a' is worse for q than a'' is) – or, we can say, that a' *harms* q (that is, that, a' is worse for q than a'' is) or that a' *thwarts an interest* that q has (since q has an interest in the agent's choosing a'' over a'). In short, the comparison between a' and a'' casts a' in a more discriminating light and thereby enables us to appreciate the fact that a', though better for p than a is, and even arguably better for q than a is, *nonetheless comes with its own still weightier moral deficiencies* – deficiencies that a, as it happens, does not share.

3.6.6 Of course, q never exists under a. An implication of my claim, then, is that agents who make choices at a given world do have obligations in respect of persons who count as merely possible relative to that world. q does, that is, exert a certain "moral pull" on the agent, first and more obviously, *not* to choose a' (the loss q incurs under a' counting against a'), and, second and in a roundabout way, *to* choose a over a' (since choosing a over a' would avoids the loss that q incurs under a').

This approach rejects Exclusion. But it *does not commit us to Inclusion*. Thus, we can say that q's plight under *a′ counts against a′* without taking the further position that q's plight under *a counts against a*. That *q* exerts a "moral pull" on the agent not to choose *a′* over *a″* does not mean that *q* also exerts a "moral pull" on the agent *to choose a′ or a″ over a* – that is, *to bring q into an existence worth having*.

Variabilism allows us to have things both ways, by taking the position that any such additional person *q* – or indeed any person at all – matters morally but also matters *variably*, depending on where the loss under scrutiny is incurred in relation to the person who incurs it. According to Variabilism, the loss *q* incurs under *a′* compared to *a″ counts against a′* but the loss *q* incurs under *a* compared to *a′* or *a″* *does not count against a*.

The ground for making this distinction between the loss that *q* incurs under *a* and the loss *q* incurs under *a′* is both intuitive and principled. The loss *q* incurs under *a′* is *morally significant* in virtue of the fact that *q does or will exist under a′*, but the loss *q* incurs under *a* has *no moral significance whatsoever*, not even the least bit, for purposes of determining whether *a′* is better than *a*, or for determining the permissibility of *a* or *a′* or any other alternative, since *q does not and will never ever exist at all under a*.

Variabilism thus describes a middle ground between Exclusion and Inclusion. We need not say that *only actual persons matter morally* – that *only their losses* can have moral significance. Nor need we say that *merely possible persons* – those "voices from another world" – that beseech us to bring them into existence must be heeded. They may beseech, but we need not heed. Instead, we can say that losses incurred by any person – you, me or the merely possible – matter morally, *if and only if* that loss is incurred in a circumstance (at a world) where the person who incurs that loss does or will exist. q's moral status is thus *variable*. The fact that *q* suffers a loss under *a′*, as compared to *a″*, does at least *bear on* our evaluation of *a′*, but that *q* suffers a loss under *a*, compared to *a′* or *a″*, *cannot bear on* our evaluation of *a* at all.

3.6.7 Variabilism in hand, we can now articulate an objection against McMahan's argument that Pill 1 is impermissible in Scenario II.

He wants to say that, since Pill 2 is better for the early fetus than Pill 1 is, and worse for no one at all (including the pregnant woman) than Pill 1 is, Pill 2 is *morally better* than Pill 1, and that Pill 1 is itself accordingly impermissible.

It does seem that if Pill 2 truly is *morally better* than Pill 1 is, we should be entitled to infer that Pill 1 is wrong. At least, we do not yet have any reason to think that the Simple Permissibility Principle is false.

The problem with the argument is the inference to the result that Pill 2 is morally better than Pill 1 is. The Standard Pareto Principle licenses that inference *only if* the options to be compared are *same-people* options. But it is hard to construe the choices that McMahan needs to compare as same-people choices – or even as same-individual choices. The *early fetus* exists under both Pill 1 and Pill 2. The *pregnant woman* exists under both choices. And others do as well. The problem is that *the child* does not. The child obviously exists under Pill 2 – exists and suffers. But there is no basis for saying that *the child* exists under Pill 1 as well. Indeed, the very point of the early abortion is to keep *the child* – that *person* – from ever coming into existence to begin with.[143]

[143] McMahan may insist that Pill 1 and Pill 2 are same-people choices, and hence that the Standard Pareto Principle grounds the inference to the result that Pill 2 is morally better than Pill 1 is. Thus he suggests that the fetus is identical to the later person. So the child is not an additional person; rather, the child came into existence just when the fetus did. For reasons already discussed, I prefer a more intuitive metaphysics. See part 1.5.4 above. But if we must adopt McMahan's, we should keep in mind that a different metaphysics is not going to erase the substantive moral issue. It would force us to *rewrite* the debate, but it would not make the debate go away.

To see this, we must, briefly, examine McMahan's metaphysics. The "fetus is the same individual as the person into whom it might later develop." McMahan (2006), p. 627. This is, he says, a "widely accepted assumption" that he himself makes for purposes of constructing the Abortion Paradox. McMahan (2006), p. 626. I have here assumed that our subject is the *early* fetus. It is otherwise hard to see how McMahan justifies claim (1) of Scenario I – that is, the claim that Pill 1 is permissible. But if our subject is the early fetus, I think it is extremely unclear whether the metaphysical assumption is "widely accepted."

But let's suppose that the early fetus is identical to the later person. We have, in that case, a biological organism that, for awhile, is a non-person and then becomes a person. It's true that, if we are bound by that view, then my *articulation* of my objection against the Abortion Paradox will fail. But it will not follow that the *substance* of my objection will fail as well. It would just mean that a great deal of re-articulation will have to take place. One option would be to rewrite Pareto Plus to reflect the idea that while the same *individuals* exist under the two choices, an *additional person-phase* – though not an additional *person* – exists under one but not the other. My objection would then proceed just as it proceeds against the original version of Pareto Plus. As long as the case is same-person-phase, we can rely on the innocuous (but rewritten) Standard Pareto Principle. The moment we must address an additional-person-phase case, however, we must rely on the highly suspect (but now rewritten) Pareto Plus.

More generally, if the metaphysics that McMahan proposes turns out to be correct, then the debate here will need to be rearticulated. Instead of talking about when a person comes into existence, we will talk in terms of when the live (human or non-human) organism transitions from its *non-person-phase* to its *person-phase*. And we will need to revise the definition of *early abortion*. It will become, not the choice not to bring a person into existence, but rather the choice not to allow the live organism to transition to its "person-phase." And the issue of moral status will become, not whether we have obligations in respect of merely possible *persons*, but rather whether we have obligations in respect of individuals who have not yet transitioned and never will transition to their respective *person-phases*.

There is thus nothing to be gained by insisting on an identity between the early fetus and the later person. As Earl Conee puts the point, the "metaphysical study of persons exhibits no moral bias." See generally Conee (1999), pp. 619–646. In other words, no question against McMahan's argument to paradox is begged by our taking it for granted here that the early fetus and the child are distinct, and that the child exists under the fetal injury choice but not under the early abortion choice. But a different metaphysics is not going to make the substantive moral debate go away. For a clear account of the "irrelevance of metaphysics to the moral issue," see Conee (1999), pp. 619–646.

Now, one might think that fact doesn't change the analysis in any material way. One might think Pareto Plus serves our purposes just as well. After all, by hypothesis, the child – the additional person – does have a life worth living – and we concede that Pill 2, accordingly, does not make things worse for that child, and even makes them better, than Pill 1 does. All other facts remain the same – including the critical fact that Pill 2 is better for the fetus than Pill 1 is. We now infer, citing Pareto Plus, that Pill 2 is morally better for the fetus than Pill 1 is.

The Standard Pareto Principle is perfectly plausible. But Pareto Plus is clearly problematic. The mere addition to an option of a person whose life is worth living cannot be counted on *not* to make things worse. It's true that Pill 2 makes things no worse, and perhaps even makes things better, for the child than Pill 1 does. But that fact cannot obliterate *just how bad Pill 2 makes things for the child* – compared, of course, not against Pill 1 but rather against No Pill.

Pareto Plus, in short, is incapable of taking into account, when comparing Pill 1 against Pill 2, *just how bad Pill 2 is.* It blinds itself to an important loss that the child incurs under Pill 2. So – of course – it is not reliable on the issue of whether Pill 2 is morally better than Pill 1 or, more critically, the issue of whether Pill 1 is itself permissible.

3.6.8 Variabilism, in contrast, considers that loss to have moral significance, both for purposes of evaluating Pill 2, which imposes that loss, and for purposes of evaluating each alternative to Pill 2, including Pill 1. That loss thus counts against Pill 2 – and, in a roundabout way, counts in favor of Pill 1 in virtue of the fact that Pill 1 itself avoids imposing that loss on the child. It is a loss that must be taken into account, according to Variabilism, in deciding the permissibility of Pill 1 and Pill 2 and – if indirectly – must be taken into account in determining whether Pill 2 truly is morally better than Pill 1 is.

We can now see just why it makes so much sense to say – as McMahan himself was willing to do in analyzing Scenario I – that Pill 1 is permissible even though Pill 2 is impermissible and even though Pill 2 meets the conditions for betterness that Pareto Plus establishes. In point of fact, those conditions are insufficient to establish moral betterness; and the fact that Pareto Plus implies that a particular

---

Moreover, the metaphysics I am assuming for purposes here is highly intuitive. It is at least plausible that the early fetus is distinct from the *person* – the kind of being whose coming into existence is signaled by that first thought. After all, *I* am a person, and *I* do not continue to exist when I have been lost to a permanent vegetative coma. Instead, *I* cease to exist at whatever point it is at which I happen to have that last thought. It potentially confuses things to say that I *am identical to* that barely alive human body. We *could* talk in terms of person-phases and non-cognitive-and-barely-alive-human-body-phrases and corpse-phases. But dividing the world up into distinct things in a natural and intuitive way, to facilitate our articulation of the natural and intuitive point that it is the *corpse* that is cremated or buried and *not me*, has its conveniences.

The point seems just as applicable to the beginning of life as to the end. The biological organism – the non-cognitive but live human body – begins to exist early in the pregnancy. *I*, however, do not enter the picture until much later, until (at least) that first thought. On this way of looking at things, while the person and the biological organism overlap in space for a period of time, they do *not* perfectly coincide in time. And they are not numerically identical.

wrong choice is morally better than a second choice puts no pressure on us at all to say that that second choice must be wrong as well.[144] For Pareto Plus is false.

3.6.9 McMahan needs more than the Standard Pareto Principle to reach the result that Pill 2 is better than Pill 1. He needs *Pareto Plus*. In fact, however, we cannot be sure that the addition of a person whose wellbeing is made to decline in some *avoidable* and *unjustifiable* way *cannot* make things *worse*. If we think that it can, then we must reject McMahan's argument in favor of claim (1′) – in favor, that is, of the claim that Pill 1 is impermissible in Scenario II.

Variabilism thus insists that the determination of whether Pill 2 is morally better than Pill 1, and the permissibility of Pill 1, must take into account how well the child will fare under No Pill. The loss that child incurs under Pill 2 has moral significance and must be taken into account.

That will not mean, however, according to Variabilism, that the losses of that same child will have moral significance *across the board*. Thus, the loss the child incurs under Pill 1, according to Variabilism, has no moral significance at all for purposes of evaluating Pill 1 – or any of its alternatives. The fact that Pill 1 means that the child will never exist at all cannot itself be counted *against Pill 1*.

Finally, in combination with the otherwise plausible permissibility theory – and here we can cite OPPP4 – Variabilism fully supports the view that Pill 2 is wrong. The loss the child incurs under Pill 2 is very much imposed in circumstances in which the child itself does or will exist – and, accordingly, counts very heavily against Pill 2.

3.7 Loss, Variabilism and Pareto Plus

3.7.1 We think that the child's loss under Pill 2 – the fetal injury choice – *bears on* the evaluation of Pill 2. After all, it is the fact that the No Pill choice *avoids* that loss without imposing too great a loss on anyone else that grounds our view that Pill 2 is wrong. We also think that the child's loss under Pill 2 bears on the evalua-

[144] See also his *Pareto Paradox*, where he makes exactly the same point. Pill 2, we agree, is impermissible. But "[P]ill 2 is … Pareto superior to Pill 1. How, then, could it be true [that] the woman may take Pill 1 but not Pill 2?" McMahan (2006), p. 641. How can a choice that is Pareto-inferior to a choice that is clearly wrong itself be permissible? McMahan himself, however, stands by the claim that it is – in both Scenario 1 and in the Pareto paradox. We should not think that the Pareto-inferiority of Pill 1 to Pill 2, together with the impermissibility of Pill 2, entails that Pill 1 is impermissible. And we can reject that inference, he writes, in virtue of the fact that it "excludes the direct comparison between Pill 1 and No Pill" – that is, the very comparison that led us to conclude (according to McMahan) that Pill 1 was permissible in Scenario I to begin with.

But I think the better solution to the Pareto paradox is one that explains just why what McMahan calls the Pareto superiority of Pill 2 to Pill 1 – which, to avoid confusion, we should instead call the Pareto-*plus* superiority of Pill 2 to Pill 1 – fails to imply *moral* superiority. We can agree, that is, that Pill 2 is Pareto-plus superior to Pill 1. But we can and should say, as well, that it does not follow that Pill 2 is morally superior to Pill 1.

tion of Pill 1, in a roundabout way. By counting *against* Pill 2, it counts *in favor of* Pill 1. More specifically, it is what shows that Pill 2 is itself *too defective, too dreadful*, to show that Pill 1 is wrong. It is what shows that Pill 2 is not morally better than Pill 1 is. At the same time, we also think that the child's loss under Pill 1 does *not* bear on the evaluation of Pill 1.

These thoughts together might seem inconsistent. But if that is so, the reason they seem inconsistent is that we are thinking about *persons* and whether *they matter morally*. *Of course* the child matters morally; that's why the child's loss under Pill 2 bears on the evaluations of Pill 1 and Pill 2. But the child does not *suddenly stop mattering morally* when we move to evaluate Pill 1. And so we seem forced to agree that the child's loss under Pill 1 bears on the evaluation of Pill 1.

3.7.2 More generally, it confuses our analysis, or at least makes things more complicated, to focus on *persons*. We might then make the mistake of thinking that, since it clearly matters morally that the child suffers a loss under Pill 2, then it must clearly matter morally that *that same child* suffers a loss under Pill 1.

Variabilism suggests that we instead focus on *loss*. We can then say that the child's *loss under Pill 2* bears on the evaluations of Pill 1 and Pill 2 – without even hinting that *that same child's loss under Pill 1* bears on the evaluation of Pill 1.

3.7.3 In contrast, Pareto Plus blinds us to the child's loss under Pill 2 for purposes of evaluating Pill 1. Once that highly salient moral fact is shunted off to the side, Pareto Plus presumptuously says that Pill 2 is *morally better* than Pill 1 without first appreciating *how bad Pill 2 really is*. The Simple Permissibility Principle then steps in to declare Pill 1 wrong. (The Multifactorial Permissibility Principle, perhaps, makes things still worse, at least from a conceptual point of view: it declares Pill 1 wrong in Scenario II but not Scenario I.)

Variabilism, in contrast, recognizes that a *correct* evaluation of Pill 1 can't blind itself to loss the child incurs under Pill 2. The loss that the child incurs under Pill 2 – the loss that Variabilism recognizes as morally significant – gives us ample reason to challenge the idea that Pill 2 is morally better than Pill 1. And that, in turn, gives us ample reason to challenge Pareto Plus.

3.8 The Standard Pareto Principle, Pareto Plus and OPPP1

3.8.1 We have introduced three Pareto principles. The first of the three was just OPPP1. To repeat it here:

Otherwise Plausible Permissibility Principle 1 (*OPPP1*).

An act *a* performed at a world *w* is wrong *if*:

(i) Act *a* imposes a morally significant loss on a person *p* at *w*;
(ii) An act *a′* at an accessible world *w′* performed in lieu of *a* at *w* is better for *p* at *w′* than *a* is for *p* at *w*; and
(iii) Either (a) *a′* imposes no morally significant loss on any person *q* at *w′* other than perhaps *p*, or (b) if *a′* imposes a morally significant loss on a person *q* at

w', then a imposes a morally significant loss on q at w, and a' is at least as good for q at w' as a is for q at w.

The Variabilist reading of OPPP1 generates all of the results (and more) that we obtain from the Standard Pareto Principle in combination with the Simple Permissibility Principle. OPPP1 in combination with Variabilism does not, however, duplicate the problematic results we obtain from Pareto Plus in combination with the Simple Permissibility Principle. If an additional person q incurs a loss under a' (as compared to how that person q fares under some third alternative a''), that loss will be deemed morally significant by Variabilism. That means that the conditions that OPPP1 establishes – in particular, clause (iii) – will not be met; and the inference under OPPP1 to the result that act a is wrong will be blocked.

3.8.2 One final point. In some important cases, OPPP1, in combination with Variabilism, *goes beyond* the Standard Pareto Principle. The Standard Pareto Principle is limited to same-people cases. But OPPP1 is not. OPPP1 can be triggered even when the way things are made better for one person p is by never bringing still other persons into existence to begin with. OPPP1, accordingly, could be renamed *Pareto Minus*.

Thus, if the alternative act a' fails to bring into existence some of the people who do or will exist under a, but makes still others – p – better off, and still manages to avoid imposing a morally significant loss on *anyone else at all* (existing, future or merely possible, relative to either w or w'), then OPPP1 in combination with Variabilism implies that a is wrong. That's a pretty stringent condition! But there is nothing in it that says that, for a to be declared wrong, each person who does or will exist in w *must also exist in* w'. The requirement is, rather, that a′ must impose *no morally significant loss at* w' *on anyone at all* (other than perhaps p).

It's true that, under the maximizing account of loss we have adopted here, any person that a′ leaves out of an existence worth having incurs a *loss* at w′. But that loss, according to Variabilism, is devoid of moral significance. Thus: the choice not to leave persons out of existence – the choice to bring them into existence – *can be wrong* even when that choice makes things much better for those persons and avoids a loss on their behalf, and even when that choice makes things only slightly worse for still others. It will be *those others* who have incurred a morally significant loss that could have been avoided *without* imposing a morally significant loss on anyone else at all. Their loss is what makes the act in that case wrong.

Chapter 4
Three More Arguments Against Early Abortion

4.1 Introduction

The main purpose of this chapter is to examine three arguments – from Hare, Marquis and Harman – that, whether by design or not, place the permissibility of the early abortion in doubt. Each of the three arguments is grounded in positions that seem to assign at least some moral significance to the loss incurred by a merely possible person when that person is left out of existence altogether. Those positions are, in turn, at odds with the account of the moral significance of loss that Variabilism itself suggests.

4.2 Variabilism and the Timing of the Abortion

4.2.1 I begin with a brief sketch of Variabilism's treatment of loss in the context of abortion. A fuller range of issues will be addressed in Chapter 5. But a brief sketch is in order here for purposes of clarifying the distinction between Variabilism and the three alternative positions I focus on in this Chapter 4.

According to Variabilism, it is critical *just when*, during the process of gamete production, conception, implantation, fruition, birth and early rearing, a person comes into existence. In particular, whether that process is interrupted *before* a person comes into existence or *after* a person comes into existence will often determine the permissibility of the abortion choice itself.

Timing is important, according to Variabilism, since whether the loss imposed by abortion is morally significant – whether that loss *bears on* the permissibility of the choice of abortion and its alternatives – depends on just *where that loss is incurred in relation to the person who incurs it*. And that fact is itself a function of timing. Thus, the loss incurred by a person *p* will have *no moral significance at all* if the abortion – or non-conception or non-implantation – takes place *prior* to p's own coming into existence, since in that case the loss will be incurred at a world where *p* never exists at all. And the loss incurred by *p* will have *full moral significance* if the abortion – or infanticide – takes place *after p* comes into existence, since in that case the loss that *p* incurs is incurred at a world where *p* exists.

M.A. Roberts, *Abortion and the Moral Significance of Merely Possible Persons: Finding Middle Ground in Hard Cases*, Philosophy and Medicine 107,
DOI 10.1007/978-90-481-3792-3_4, © Springer Science+Business Media B.V. 2010

4.2.2 Variabilism itself simply determines what losses are morally significant and what losses are not. On the basis of that moral data, the otherwise plausible permissibility theory – which plausibly includes OPPP1-OPPP4 – then evaluates the abortion choice.

If the loss the person *p* incurs is *not* morally significant, then the otherwise plausible permissibility theory will not count that loss against the choice of abortion or, in a roundabout way, in favor of any alternative. But if the loss *p* incurs *is* a morally significant loss, then the otherwise plausible permissibility theory will count that loss against the abortion and in favor of its alternatives.

4.2.3 Let's first take the case where the loss incurred as a result of the abortion is incurred by a person who never comes into existence to begin with. Here we are talking about the *early abortion* – the abortion (relative to the world of performance) that relegates a particular person to the class of the merely possible. According to Variabilism, that loss is devoid of moral significance.

With that loss out of the picture, in any ordinary case, the otherwise plausible permissibility theory can be expected to imply that the early abortion is *permissible* – as long as it is what the pregnant woman wants – whether it is done for a good reason, a poor reason or no reason at all.

Now, there will be exceptions to this general principle. Conflicts will sometimes arise among persons other than the person who, as a result of the early abortion, never comes into existence to begin with. And the losses such persons face in many cases are losses Variabilism will deem morally significant. Thus, it may be in the interest of the pregnant woman herself to have the abortion and in the interest of the man who impregnated her – or the woman's grandmother, or still others – that the woman continue the pregnancy. Then and only then does the analysis become particularly complex. For then and only then do we have a situation in which the only way to avoid causing one person to incur a morally significant loss is to impose a morally significant loss on someone else. Just how the tradeoff is to be made will, in some cases, be resolved by OPPP4 – and, more generally, by a more complete version of the otherwise plausible permissibility theory.

Still, we can anticipate that the applicable tradeoff principles will surely resolve many such conflicts in favor of the pregnant woman. After all, she ordinarily has the most at stake. The morally significant loss that she will incur if the early abortion is not performed ordinarily will be considerably greater and deeper than any *morally significant* loss anyone else will incur if the early abortion is performed.

Moreover, in many cases, there will be a way short of requiring the woman to continue a pregnancy she doesn't want to compensate for any deep loss the man – or the grandmother or still others – may incur. The man may choose another partner, and the woman's grandmother may be willing to accept that a great-grandchild will come *later*.

Finally, at least in some cases what seems like a loss will not be a *genuine* loss at all, but rather a *sense* or *expectation* of loss. While the *sense* of loss may well make for a loss of sorts, it is also likely that the person who has that sense of loss has, or can be expected to locate, the resources necessary to mitigate it – the inner strength; the help of a friend; professional counseling. It is likely, in other words, that there will be some very good way of dealing with any such sense of loss that is consistent with the pregnant woman's proceeding with the early abortion.

4.2.4 We now turn to the case where the loss incurred as a result of the abortion is incurred by an already-existing person – the case, that is, where the abortion involves ending the life of an existing person. Here, we are talking about the *late abortion*. And here the loss incurred is, according to Variabilism, a morally significant loss.

We can anticipate that the otherwise plausible permissibility theory will now analyze things quite differently. It's true that the loss the early abortion will impose is just as great and deep as the loss that the late abortion will impose. But, as we have already seen, the loss imposed by the early abortion, according to Variabilism, is devoid of moral significance. In contrast, according to Variabilism, the loss imposed by the late abortion on the new person *p* has full moral significance. –So, of course, does the pregnant woman's. Still – unless we are in a case where *p*'s life will unavoidably be less than worth living – it is very likely that the loss *p* faces will be greater and deeper than any morally significant loss that anyone else will face. And we can accordingly expect the otherwise plausible permissibility theory to resolve the conflict in favor of *p*.

It is thus critical, according to Variabilism, *just when* during the process of gamete production, conception, implantation, fruition, birth and early rearing a *person p comes into existence*. When, during the process of childmaking, is it that we actually have a *person* rather than something that is a live human organism but *not* a person?

4.2.5 For purposes of this book, I accept as an assumption what I earlier called the Thinking Thing Account of when existence commences in the case of persons.[145] According to that account, a pregnancy does not involve a *person* – the kind of thing, that is, in respect of which we have moral obligations; the kind of thing we must generally create more wellbeing for rather than less *at least* in the case where it exists – until the point in the pregnancy at which *thinking* itself has emerged.

And when is that? For purposes here, I do not try to answer that question in any complete way. It does seem clear that thinking itself requires more than electrical activity in the brain or a mechanical response to pain stimuli. Yet it also seems plausible that thinking often takes place beyond our own conscious awareness. Moreover, *intermittent* thinking alone is enough to keep one in existence. Once in existence, a person can continue in existence even if that person is not *continuously* thinking. The person doesn't cease to exist, then, until the conclusion of that last thought – even if that person's body, sustained by natural or artificial means, clings to life far beyond that point.

On this way of looking at things, the embryo and the early fetus are *live human organisms*. But, until thinking has emerged, there is no *person* there. Since *person* itself is simply defined in moral terms, the most contentious point here is that *persons* are the kinds of things whose coming into existence is *signaled by their thinking*. The most contentious point here, in other words, is that being a *thinking thing* and *mattering morally* – being, that is, the kind of thing in respect of which

[145] See part 1.6 above. See also Chapter 5.

we have obligations; being the kind of thing we must create more wellbeing for rather than less *at least* if it exists – come together.

I will not try to say *why* that is so here. Plausibly, however, the reason that being a thinking thing and mattering morally come together is that it is in the case of thinking things that it *matters to them* whether we treat them in one way or another.

On this view, those who matter morally (if variably) and those who don't matter morally (not even variably) – persons and non-persons – are not divided in accordance with their species membership. Some individuals who are arguably human count as non-persons, in other words, and some non-humans count as persons.

4.2.6 Putting the view just described together with Variabilism, we can say that, if the biological organisms that constituted our own early bodies never had had their first thoughts, then *we* never would have existed at all. And any losses imposed on us at any world where *we* never existed at all are *devoid of moral significance*. They do not count against the choices that impose those losses or, in a roundabout way, in favor of the alternatives that avoid those losses.

But once we begin to think – more precisely, once *we* come into existence; once the human organism begins to think – things are very different. For any loss we incur at any world where we do or will exist has *full moral significance*. Depriving us of our entire futures once we exist thus constitutes a *morally significant loss* – and in many cases it will constitute a greater and deeper loss than the loss the pregnant woman or anyone else would be forced to suffer if the abortion never takes place at all.

4.2.7 Many of the points I have just made – that the embryo and early fetus are not persons; that the late fetus and human neonate ordinarily are persons; that thinking and mattering morally come together – are adopted, for purposes here, as assumptions. But they don't beg any question. That is so, since the issues they raise are not in dispute here. Even if all of those issues happen to be resolved entirely in my favor, the issues that are in dispute here will remain quite alive. For the arguments against early abortion that I examine here are not driven by the claim that the embryo, or the early fetus, is a *person*. The authors of those arguments – Hare, Marquis and Harman, and also McMahan, – can easily *accept* the assumptions that I have made here even if they regard those assumptions in need of tweaking at the margins. What they won't accept is *just how* I have claimed the thing that is unequivocally a *person* – that is, the *possible person* that the embryo or early fetus may eventually develop into, depending on how the choice of early abortion is itself made – *matters morally*. What they won't accept is my claim that the merely possible person *matters variably*.

Thus for purposes here the open questions are these. Can the loss that a merely possible person incurs, as a result of the early abortion, make that otherwise permissible choice *wrong*? What are the implications, for our evaluation of the choice of early abortion, that any one of a vast number of *merely possible persons* could have had just as wonderful a life as I have had (to date)? What are the implications, for the evaluation of that choice, that the future any such merely possible person faces is not really particularly distinguishable from my own? What does it mean, for the evaluation of that choice, that one or more possible persons will have existences worth having if the choice of early abortion is left *unperformed*?

4.3 The Golden Rule: Hare

4.3.1 It may not be surprising that some early discussions of abortion flirt with significant problems in population ethics – including, among others, the *Repugnant Conclusion*. Surely by 1975 it was beginning to become clear just how interesting those problems in fact were. Thus, in one short paper published in just that year, R.M. Hare worried not just about abortion but also about the "general duty to produce children" – which he somewhat hesitantly endorses on a rebuttable basis – and whether, by accepting that that duty exists, he thereby commits himself to "unlimited procreation."[146] (He admits, wryly I hope, that "it is morally all right for the parents to decide, after they have had, say, 15 children, not to have any more...."[147])

But he unfortunately – in a move Parfit would question only a few short years later in *Reasons and Persons*, and that many today would reject on the basis of the Repugnant Conclusion[148] – relies on the "classical utility principle" – what I have here called *Totalism* – to set all appropriate limits. The more the merrier, in other words, until we get to that point at which the additional person "impose[s] burdens on the other [persons] great enough in sum to outweigh the advantage gained by the additional [person]. In utilitarian terms, the ... total utility principle sets a limit to population which, although higher than the average utility principle, is nevertheless a limit."[149]

4.3.2 While Hare's conclusions have the support of Totalism, his argument against abortion is in fact formulated in terms of the Golden Rule: "we should do to others as we wish them to do to us."[150] Extended a bit, the rule becomes: "we should do to others what *we are glad was* done to us."[151] If we are glad that no one "terminated the pregnancy that resulted in *our* birth, then we are enjoined not, *ceteris paribus*, to terminate any pregnancy which will result in the birth of a person having a life like ours."[152]

A like argument – Hare is the first to admit – applies to the choice not to conceive.[153] The fact that the one object – the pair of gametes that might combine to produce a child – is spread out in space and the other – the early fetus – is not does not, in his view, represent a material distinction between the two cases.

4.3.3 I agree with this last point. If the argument he presents against – say – early abortion is sound, so is the argument against non-conception. As he puts it, "[i]f I have a duty to open a certain door, and two keys are required to unlock it, it does

[146] Hare (1975), p. 218.

[147] Hare (1975), p. 211.

[148] Parfit (1987), pp. 381–90.

[149] Hare (1975), p. 218 (citing his review, Hare (1973), of Rawls, who himself was famously worried about utilitarianism's failure "to take seriously the distinction between persons." Rawls (1972), p. 27).

[150] Hare (1975), p. 208.

[151] Hare (1975), p. 208.

[152] Hare (1975), p. 208.

[153] Hare (1975), p. 212.

not seem to me to make any difference to my duty that one key is already in the lock and the other in my trousers."[154]

I do, however, object to his particular extension of the Golden Rule. From the fact that I am glad that no one interfered with my bank heist, it does not follow, not even *ceteris paribus*, that I am obligated not to interfere with the bank heists of others. It may be hypocritical for me to do so, but it isn't necessarily morally wrong. Similarly, I am thrilled to pieces that my parents decided to have a child soon before they conceived me and that my mother did not then change her mind and opt for abortion. But how we infer from that fact that I am obligated *not* to interfere with the coming into existence of still others is a mystery.

A more plausible principle would include a moral evaluation of some sort in its antecedent. An example of such a principle would be this: if our parents had an obligation not to interfere with our coming into existence in virtue of the fact that we are glad that we exist, then we are obligated not to interfere with the coming into existence of others in virtue of the fact that they will be glad that they exist. If we take the view that *they* were obligated to conceive and not to abort us in virtue of the fact that that is what is good for us, then we may be compelled to say that we have an obligation to conceive and not to abort still others in virtue of the fact that that is what is good for them. The original principle, in other words, should itself be amended, so that it tries to tell us something about the views we are bound to take regarding what we are obligated to do, rather than something about what we are obligated to do. And such an amended principle would be powerful: if the antecedent held, the principle would certainly have important implications for the sorts of conclusions regarding non-conception and abortion we would be bound to reach here.

4.3.4 Now, it may seem that this amended principle can be challenged as well, on the ground that there is a morally relevant distinction between the class of persons who do exist (relative to a given act performed at a given world) and the class of persons who never will exist (relative to that same act and world). Hare himself considers that objection: the "logician," he writes, might counter that "these potential people do not exist … and therefore cannot be the objects of duties"; "only actual people have [rights or interests]."[155]

George Sher subsequently put the point in a more general way. Not all entities, he argued, qualify for treatment under the Golden Rule.[156] For the Golden Rule to imply that it is wrong to interfere with a new person's coming into existence, we must first establish that that new person is the sort of entity to which the Golden Rule applies. If we just assume that point, then our argument is circular.

4.3.5 Now, if the abortion is not in fact performed, and the new person *does exist*, then it would be hard to argue that that the Golden Rule would not apply to that person. But suppose the abortion is performed, and the person who might have existed is left as a *merely possible person*. In that case, Sher's objection seems to

[154] Hare (1975), p. 212.

[155] Hare (1975), p. 219.

[156] Sher (1977), p. 186.

hit home (and coincide with the concern Hare himself tries to address): it may seem at least plausible to think that there is an important moral distinction between persons who do or will exist (think Hare's *actual* persons) and merely possible persons, persons who will never in fact exist at all.

I want to suggest, however, that this way of objecting to Hare's argument will not in the end be satisfactory, for two reasons. First, as Hare himself notes, "it would be strange if there were an act whose very performance made it impossible for it to be wrong."[157] We don't make acts right by performing them or wrong by failing to perform them.[158] But there is second difficulty as well. Surely our obligations in respect of the merely possible truly are governed by something like the amended form of the Golden Rule. Thus, just as I don't want to be enslaved and take the view others are obligated not to enslave me, I am obligated myself not to enslave others. And those others I am obligated not to enslave will include the *merely possible*: it would be wrong for me to bring them into existence and then, or in order to, enslave them.

This is the point that Addition Plus (Case 2.3) makes: whatever else we say about the case, we think that the choice to bring Fen into a much lesser existence, when agents had the option of bringing Fen into a much better existence, was wrong. And that's so, even if Fen never actually exists.

We should say instead that there is some rough sense in which we are all – existing, future and merely possible – subject to the Golden Rule. We should think that others should be treated as we think we should be treated. It does not seem we are going to make much headway understanding moral law if we think one set of rules applies to us and another set of rules, a set of rules that the relieves us of any obligation we happen not to want to discharge, for example, applies to everyone else.

But it is not going to follow that the abortion itself is *wrong*. Even if we think the amended version of the Golden Rule itself seems plausible, and even if we then say that that amended principle applies to the merely possible just as well as it does to you and me, there remains the question of whether the antecedent of that principle is itself satisfied.

In its amended form, and applied to abortion, what we have is just this: *if* we think that our parents had an obligation not to interfere with our coming into existence in virtue of the fact that we are glad that we exist, then we should also think that we are obligated not to interfere with the coming into existence of others in virtue of the fact that they will be glad that they exist. The problem is that it is not at all clear that we *do* think our parents had an obligation not to interfere with our coming into existence for our sake, or because our coming into existence is good for us or because we are glad that we exist. I think that – intuitively – we *don't* think that. We – intuitively – think instead that just as we are not obligated to produce additional persons for their own sake, our parents were under no obligation to produce us for our own sake.

---

[157] Hare (1975), p. 219.

[158] If we accept Rabinowicz's Principle of Normative Invariance – as I do – we will think Hare's point here is correct. See note 61 above.

4.3.6 Variabilism helps explain why it would be a mistake to think that our parents had the obligation not to interfere with our coming into existence – or, more accurately, why, if they do have that obligation, it is grounded in something other than what is good *for us*. Leaving you (say) out of existence altogether would have, let's suppose, imposed a loss on you. According to Variabilism, however, since that loss is incurred by you at a world where you never exist at all, that loss is devoid of moral significance. *That* loss thus cannot count against the abortion choice. If that choice is otherwise permissible – if it appropriately balances the interests of others whose wellbeing might be at stake – then the loss that you yourself would have incurred as a result of that choice cannot make that choice wrong.

Of course, we here distinguish between never bringing a person into existence and removing a person from existence. Variabilism is perfectly capable of making that distinction. Removing someone from existence is to impose a loss on that person at a world where that person exists; the loss that is imposed is, thus, a morally significant loss. While the otherwise plausible permissibility theory will not condemn all such acts, it will surely condemn many.

4.3.7 One final note. Hare "abjures" as unhelpful the discussion of whether the fetus itself is a person.[159] It may be thought that a significant advantage of Hare's approach is that we need not address that question: the abortion is forbidden, as long as we can agree that it is a *person* that the pregnancy in the end will generate. That is a plus, he thinks, since *person* itself is "not a fully determinate concept" and we can argue endlessly whether the early fetus, or the later fetus, or the embryo, the *human* embryo or even the human *neonate*, counts as one.[160] I have acknowledged that point above.

On the other hand, isn't there at least some sense of the term *person* that we do clearly understand, and that is useful for purposes here? I argued earlier that we think of ourselves as persons, and of ourselves as things that begin to exist when we begin to think and things that cease to exist when we cease to think. With this sense of the term firmly in mind, we can plausibly take the position that persons are also exactly the kinds of things that matter morally – at least if they exist – perhaps simply in virtue of the fact that it *matters to them* how we treat them. So, if we think that our task is that of giving the one, true definition of the term "person," then we should give up. But there is no reason that we need to consider that our task.

4.4 Futures of Value: Marquis

4.4.1 In both his highly regarded 1989 paper arguing against abortion and in a more recent paper, Don Marquis aims to (i) reinvigorate the intuition motivating Hare's argument, (ii) avoid, as Hare does, any need to say just when during the process of

159 Hare (1975), p. 202.

160 Hare (1975), pp. 205–206.

conception and fruition a *person* comes into being and (iii) avoid, as Hare does *not*, a commitment to what seems an implausibly oppressive procreation obligation.[161]

4.4.2 Marquis begins with a general account of why death is often a misfortune and why killing is "presumptively" wrong – wrong, that is, in the absence of adequate justification. On Marquis's view, most of us have before us a *future of value*, a future that, however it precisely unfolds, will include many good things. If we are killed now, we lose that future of value – we lose something that is *of value to us*. For that reason, to impose that loss on us – to kill us – is, in the absence of justification, a serious wrong.[162]

Marquis's view of death is consistent with the idea that whether a particular death is bad at all, and, if it is, how bad, can be determined by comparing how that person fares at a world where he or she continues to live against how that person fares at a world where he or she does not continue to live.[163] Of course, there are many different possible futures that may be open to us at any given time, depending, among other things, on the choices that we and others happen to make. Some of those possible futures will be wonderful for us, and some will be at least nice and a few may well be *less* than worth having. The comparison between each of these possible futures in which we do continue to exist against a possible future in which we do not continue to exist will yield an array of particular values – measures of just how much better (or in a few cases worse) for us it is that we continue to exist than that we not. As long as the value that is correlated with a particular future that includes us falls in the positive range (as long as the life is not itself *wrongful*), the death is, relative to that future, a bad thing for us. Since it is *very likely* that, whatever future in fact unfolds for me, that future will be wonderful or at least nice for me, it is *very likely*, as well, that death now would be a bad thing for me. And it would be bad for me in a sense that is relevant to whether the act that causes it is permissible. Death isn't just a *loss*. It's a *morally significant* loss.

4.4.3 This account of why death is often bad seems reasonably uncontroversial – and is consistent with Marquis's own. And it seems roughly to explain, as he puts it, why killing is presumptively wrong.

What happens next is less plausible. For Marquis then extends his account of the badness of death for the purpose of grounding an argument against early abortion.

Suppose that we do find this account of the badness of death reasonably uncontroversial. I think we are then surely making a certain critical background assumption regarding just when it is to be applied. We are assuming that the account applies – generates the result, that is, that the death is bad – just in the case where there

---

[161] Marquis (1989), pp. 183–202, and Marquis (2008), pp. 422–426.

[162] Marquis (2008), p. 422.

[163] The account of why death is often bad that follows this note in the text roughly comports with Marquis's view but isn't drawn directly from it. Rather, it derives more directly (but with various adjustments of my own) from Feldman (1994). For a brief account of the loss of death as something that is incurred by a person who does or will exist, see part 2.2.3 above.

exists a person whose future of value is eliminated by that death. We are assuming, that is, that the account generates the result that death is bad only when (1) the thing whose future of value we strip away is a *person* and (2) that thing *exists*.

In order to extend the account of the badness of death to the case of early abortion, Marquis must revise (1). But he can at the same time – and this is why his account diverges from Hare's – leave (2) as it stands.

The conditions he thus adopts on when the stripping away of a thing's future of value – when the killing of that thing – is bad and presumptively wrong are just these: (1′) that thing is an *individual* – it need not be, or in any event need not *yet* be, a *person* – and (2) that thing *exists*.

Now, not *all* individuals have a future of value that can be stripped away from them. What sorts of individuals can expect – provided that they continue in existence at all – to have futures like ours, that is, futures of value? Daffodils and ticks we can perhaps rule out; cats and dogs we can leave unjudged. The individuals that most clearly will have futures of value are, of course, human embryos and early human fetuses. They look forward, not to the future that a daffodil or a tick or even a cat will have, but to the future that individuals who are uncontroversially *persons* look forward to. They look forward to futures *like ours*, futures of *value*.

Marquis thus concludes that it is bad and presumptively wrong to end the life of a human embryo or early human fetus. The early abortion is, like any ordinary killing of any ordinary person, bad and presumptively wrong.

4.4.4 Marquis thus nowhere claims that the human embryo or early human fetus is a person. He can easily live with the assumptions that I make starting out – in particular, that the embryo and early fetus are not persons in virtue of the fact that they have not had that first thought. Marquis's argument instead proceeds from the more modest idea that the embryo and human fetus are *existing individuals* that have futures of value. And that's all – he thinks – that he needs.

4.4.5 It is Marquis's adoption of condition (2) – that the individual whose future value would be eliminated by abortion in fact *exists*; what I will call the *existence condition* – that allows Marquis to circumvent the stringent procreative obligation that Hare himself seems committed to. It is that condition that blocks the extension of Marquis's own argument to the case of non-conception. The idea – and we will look more closely at this idea below – is that until the individual who has the future of value itself *exists*, there is nothing *there* to take anything away *from*.

While Hare seems to insist that we conceive a child *and* not abort that child, what Marquis says is perfectly consistent with the idea that non-conception is permissible. It's just that, on Marquis's view, when we do conceive – when we do bring the embryo or early fetus into existence – we must not then abort it.

4.4.6 The beauty of Marquis's argument against early abortion is its facial neatness. But beneath the neatness things are a little messy. Conditions (1′) and (2) both require a very close look.

4.4.7 Condition (1′) asserts that a thing must be an *individual* in order for the elimination of its future of value to count as bad and presumptively wrong. Marquis uses the concept of the individual to argue that his account has no application to the human embryo in its very earliest stages. I take it that that concept would similarly rule out the application of his account to the scattered object consisting of just egg and sperm. That thing, after all, has every bit as much of a future of value as does the early embryo. But let's start here with the early embryo.

Until approximately the 14th day after conception, when the primitive streak emerges, the embryo itself is capable of dividing into two genetically identical (monozygotic) successor embryos.[164] It is capable of *twinning*. Marquis considers – but seems to reject – the view that the original embryo ceases to exist at the moment the twinning takes place.[165] The following argument might lead one to think otherwise. If one of the successor embryos is identical to the original, then, since the other embryo occupies the same position relative to the original, the other embryo must be identical to the original as well. But by the transitivity and symmetry of identity, we cannot consistently say that any *two* things are *identical* to any one. The argument concludes that the original embryo must not have *survived as* either of the two successor embryos; instead, it has ceased to exist at all, and been replaced by the two embryos.

The glitch in this approach that leads Marquis to reject it is the case where that identical original embryo does *not* in fact have a fission in its future.[166] In that case, it becomes implausible to say that that embryo ceases to exist. But it also seems that what we say about the embryo in the one case we should say about the embryo in the other: if it survives in the case where twinning doesn't take place, then it survives as well in the case where twinning does take place.

An alternate picture of things may accordingly seem more tenable: that, whether the embryo twins or not, the two, or four, or eight, genetically identical cells that constitute that original embryo are themselves individuals, each of whom has its own future of value. On this view, twinning does not involve the elimination of a single individual, or the survival of one individual as two or more individuals. Rather, twinning involves the survival of two or more individuals, from one moment to the next, as two or more individuals.[167]

What keeps Marquis from embracing this way of looking at things is his sense that, in the case where the twinning does *not* in fact take place, we are compelled to say that *only one individual human being* can emerge from the process.[168]

Facing these questions, Marquis seems to conclude that we should delay saying that we are dealing with an *individual* – and hence delay application of the future of value account – until the possibility of twinning is itself passed – until, that is, the

---

[164] For a clear and very precise discussion, see Louis M. Guenin (2008), p. 59 ff.

[165] Marquis (2008), p. 425.

[166] Marquis (2008), pp. 425–426.

[167] David Lewis presented something like this view in his defense of the psychological continuity account of personal identity against the brain splitting paradoxes. He imagined the two surviving persons prior to twinning to simply overlap in space for some period of time and then – if the twinning in fact takes place – going on to enjoy their separate lives. See Lewis (1970), pp. 17–40.

[168] Marquis (2008), p. 426.

14th day after conception.[169] "Whatever assumption we make concerning the first phase of a human being, the future of value analysis of the wrongness of killing does *not* support an objection to human embryonic stem cell research."[170]

4.4.8 The difficulty with Marquis's position I want to focus on here is not that he wants to say that the embryo is not yet *a single individual* until the *multiplicity* of *potential individuals* have knitted themselves together into one unravelable thing (my metaphor, not his). The difficulty isn't that he wants to say some things are individuals and others aren't. The difficulty isn't that he doesn't want to count the 5 day embryo as an individual, or alternatively as many individuals, but does want to count the 14 day embryo as an individual. At least, that's not an *obvious* difficulty (though defining *individual* is never going to be a simple task).

The difficulty I want to focus on here is the *moral weight* that Marquis *sotto voce* attaches to the term *individual*. He wants to say that the 14 day embryo is a single individual having a future of value that we are enjoined from eliminating. He wants to say that the 5 day embryo is not an individual and not a multiplicity of individuals, and hence that we are not enjoined from eliminating whatever futures of value that would otherwise be theirs.

In effect, he is saying that *some things*, that is, *individuals*, have futures of values we are enjoined from eliminating and *other things*, that is, *non-individuals*, do not. He is committing himself – without explicitly saying so – to the view, in other words, that some things, that is, *individuals*, *matter morally* and others, that is, *non-individuals*, do not.

4.4.9 The same tactic, it seems, would be deployed in the case of the *barely* scattered object consisting of a single human egg cell and single sperm cell residing in a single glass dish. Fertilization, let's suppose, is imminent. To avoid the view that it would be wrong to interfere with conception in this case, Marquis could take the position that that barely scattered object is not itself an *individual*. His account then would be consistent with the view that there is nothing wrong with our interfering with conception – and thereby depriving that scattered object of its future of value – by way of, saying, grabbing the sperm by its tail and insisting on its having its own dish. We've stripped a *thing*, but not an *individual*, of its future of value. So what we have done is not necessarily wrong.

Or recall Hare's example of the two keys and the lock he has a "duty" to open. Marquis would presumably deny that *any arbitrary pair* of egg and sperm, from *any arbitrary pair* of potential gamete providers, constitutes an *individual*. Hence, though any such pair will have a *future of value* in any scenario in which the relevant gamete

---

[169] Cynthia Cohen has recently argued in favor of the "14 day or later rule" as offering the best account of when human embryos begin to have the same moral significance as living human beings. Among other things, such a rule suggests that even if the earlier embryo is itself owed some form of special respect, stem cell research may proceed in view of the fact that the embryonic stem cells themselves would be taken from the blastocyst, and the embryo at that point destroyed, at day five or six. She, in part, supports that rule by reference to the twinning argument. See Cohen (2007), pp. 67–69, 72–73 and 84.

[170] Marquis (2008), p. 426 (emphasis added).

providers are poised to choose the relevant manner of conception over non-conception at the relevant time, depriving that pair of its future of value, by not proceeding to conceive a child, is not wrong.

4.4.10 To address the scattered object cases, Marquis might insist that a thing is not an *individual* unless its parts are *spatiotemporally continuous*. A future of value may well be in the picture, relative to the scattered object consisting of the single egg and single sperm, whether residing in the same glass dish or in human bodies on opposite sides of the planet. But if the scattered object is not an individual, then it would not be wrong, on Marquis's account, to eliminate that future of value.

Perhaps that way of looking at things is correct. But even if it is, my objection is left unaddressed. Marquis is using the distinction between individuals and non-individuals to distinguish between cases where it is wrong to eliminate a future of value and cases where it is permissible.

More generally, let's suppose that Marquis can clearly articulate a line between individuals and non-individuals. We shall still need an explanation for why things that belong to the former class *matter morally* in a way that things that belong to the latter class do not – why, that is, it is *wrong* to deprive *individuals* of their futures of value but not wrong to deprive still other sorts of things – *non-individuals* – of their own futures of value.

4.4.11 We now turn to Marquis's second condition – condition (2). According to that condition, the thing whose future of value is eliminated – that is, the *individual* whose future of value is eliminated – must *exist*, in order for the elimination of that future of value to be considered bad or presumptively wrong. Accordingly, depriving the *merely possible* individual of its future of value does not, according to Marquis, even begin to trigger the result that that deprivation – that "killing" by way of relegating an individual to the class of the merely possible – is wrong.

Marquis thus adopts what I am calling the *existence condition*. But he doesn't fully explain it. It is way to avoid the embarrassment of Hare's overly stringent procreative obligation. But it is a way that seems at odds with certain aspects of Marquis's own account.

Thus, just as we can legitimately ask why Marquis thinks *individuals* matter morally but *non-individuals* do not, we can ask why Marquis thinks *existing* things matter morally but *merely possible* things do not. After all, the merely possible person has just as much of a future of value, immediately prior to conception, as the 14 day human embryo does. Yet, on Marquis's view, the fact that embryo has a future of value means that *the embryo* is the kind of thing in respect of which we have moral obligations: when we take its future of value away from it – when we impose that loss on it – we do something that is presumptively wrong. Why, then, don't we do something just as wrong when we interfere with conception and take the future of value away from that *merely possible person*? Why doesn't that loss, as well, matter morally? Why does the speck that consists of the human embryo in some way *inherit* the value of its possible future, while the merely possible person, according to Marquis, does not?

4.4.12 Marquis might answer that the importance of existence is simply automatic. The difficulty with that answer is that it seems at odds with the explanation he would give of why the loss of a *future of value* incurred by the *14 day embryo* has moral significance, but the loss of a *future of sorts* incurred by the daffodil or the tick does not. It seems that it is the fact that the loss is the loss of a *future of value*, and not just a loss of a *future of sorts*, that is doing the work of conferring the moral significance. It is the *nature of what stands to be lost* that seems to be doing the work.

But if that is so, then the loss the merely possible person incurs when that person is left out of existence altogether should itself have moral significance. After all: that merely possible person has just as much of a future of value as the 14 day embryo.

Of course, to block the implications of that way of looking at things for the case of the merely possible person, Marquis can – simply insist on the existence condition. But in doing so Marquis raises the question of just what we are to say about the *future of value* per se. If it is *existence* that has the power to imbue some losses but not others with moral significance, then it seems that it is not, after all, the future of value that has that power.

4.4.13 Another option – and in fact the option that Variabilism suggests – would be to say that losses of futures of value are just like any other losses. What makes them have moral significance or not is where those losses are incurred in relation to the individual – or I would say the person – who incurs them.

Variabilism thus adopts something like the existence condition, but in a carefully tailored form. According to Variabilism, whether the loss has moral significance or not depends on just where that loss is incurred in relation to the person who incurs it. Variabilism is itself controversial. But it avoids a bright line between existing, or actual, persons, and the merely possible. It instead adopts the more plausible view that they *all* matter morally, but on a *variable* basis. We can then agree that eliminating a future of value for a person who never exists at all – by never bringing that person into existence to begin with at a given world – cannot itself constitute a morally significant loss. But we also want to recognize very clearly that that same person can incur morally significant losses in abundance at still other worlds – worlds, that is, where that person does or will exist.

4.4.14 One of the reasons Marquis's account seems so neat on first inspection is that it seems not to require us to say just when during the pregnancy a *person* comes into being. The concepts he wants to work from seem far more neutral. All and only those things that are *both individuals and existing* are such that depriving them of their futures of values is bad and presumptively wrong.

The problem is we immediately want to know why it is bad and presumptively wrong to eliminate futures of value from *all and only* those things that are individuals and exist. In asking that question, we immediately open the whole can of worms. To call a 14 day embryo, but not a 5 day embryo and not an egg cell a nanosecond away from fertilization, an *individual* is not going to justify the result that it is wrong to deprive the 14 day embryo, but not the 5 day embryo or the imminent conceptus, of its future of value. We need to know more. We need to know *why* the loss incurred by the one but not the other has moral significance – why it is bad and presumptively wrong to impose the one loss, and to eliminate the one future of value, but not the other.

4.4.15 Marquis's approach is refreshing in part because it tries to find a way to avoid the arduous work of determining when it is during the pregnancy a *person* comes into being. In contrast, Variabilism accepts the need for this work. For it determines whether a particular loss has moral significance by reference to the class of *persons*: existing persons, future persons and the merely possible. For purposes here, I have assumed that *persons* are those things moral law requires us to create more wellbeing for rather than less *at least* in the case where they exist at all. We ourselves belong to that class. Moreover, the fact that we can give a plausible account of just when we come into existence and when we will go out of existence suggests that we can give a plausible account of when persons in general come into existence and when they go out of existence. I have suggested that these beginning and end points are marked by the individual's first and last thoughts.

These points are, of course, controversial. But the idea that losses bear on the moral permissibility of the acts that impose those losses and their alternatives *only if* those losses are incurred by *persons* at least *potentially* explains – in a way that calling one an *individual* and the other a *non-individual* never can – why it is wrong to deprive the 3-year-old child of its future of value but not wrong to deprive the egg cell whose fertilization is imminent of its future of value.

4.5 The Actual Future Principle: Harman

4.5.1 Hare's approach coincides with the "classical utility principle" – that is, *Totalism*. According to Totalism, agents are obligated to make the choice that maximizes wellbeing on an *aggregate* basis. If leaving a merely possible person out of existence means that less aggregate wellbeing is created rather than more, then it is wrong, according to Totalism, to leave that person out of existence. In this way, Totalism elevates what happens to merely possible persons when we leave them out of existences worth having to what happens to existing and future persons when we create less wellbeing for them when we could have created more: losses of *both* sorts are considered to have *full moral significance*.

Marquis adopts an *existence condition*. Thus on his account the loss incurred by the possible person when we leave that person out of existence altogether has full moral significance *only if* the individual human organism – the early human fetus, or even the embryo once it no longer has a potential fission in its future – that will develop into that person *in point of fact exists*. Thus, Marquis's account, unlike Hare's, avoids the implication that the failure to conceive a child to begin with, or the in vitro conception of a child followed by the failure to arrange for the 8 or 16-cell embryo to develop *in utero*, is inherently wrong.

Elizabeth Harman also adopts an existence condition for purposes of addressing certain questions regarding the moral status of the human embryo and early human fetus. The existence condition she adopts is more stringent than Marquis's in one respect and less stringent in another. According to Harman, a *person* or other being that has "intrinsic properties that themselves confer moral status" must come into existence if the abortion is to be declared wrong.[171] However, that person need not *already* be in existence in order for the early fetus to have moral status. It is enough that that person will come into existence *at some future time*. Thus, it is not just the properties a thing *now* has that are important in determining whether that thing "has moral status."[172] Rather, a "thing's present nature" – and ultimately its moral status – is "solely determined by the intrinsic properties it *ever* has."[173] This is what Harman calls the *Actual Future Principle*.

In terms of worlds, we could say that the embryo or early fetus has moral status at a world w, according to Harman, *if* the person the early fetus develops into at some world or another *w′ in fact eventually comes into existence at w.* Thus: "each fetus has its status in virtue of facts about that fetus's actual life: these facts might have been different."[174]

4.5.2 The Actual Future Principle seems to imply, among other things, that "early fetuses have their moral statuses contingently."[175] The early fetus that "will become a person has some moral status" while the "early fetus that will die while it is still an early fetus has no moral status."[176] The fetus, we can say, at one world is identical to the fetus at another. But that will not mean that the fetus has the same intrinsic properties at both worlds. According to Harman, then, "there are two significantly different kinds of early fetuses."[177]

On this view, the present nature of a particular early fetus in part depends on what happens next. At a world where there is no early abortion, we will be dealing with an early fetus that has one sort of nature and some moral status. But at a world where the early abortion does take place, we will be dealing with an early fetus that has another sort of nature and no moral status at all.

4.5.2 The Actual Future Principle enables Harman to address two issues – early abortion and how an early fetus can be an appropriate object of parental love.

[171] Harman (2000), p. 310.

[172] Harman (2000), p. 318.

[173] Harman (2000), p. 318 (emphasis added).

[174] Harman (2000), p. 320.

[175] Harman (2000), p. 321.

[176] Harman (2000), p. 311.

[177] Harman (2000), pp. 311–312.

According to Harman, the early abortion is deemed permissible – or at least a choice that requires "no moral justification" – *if* the person or conscious subject the early fetus *might* develop into *never in fact does* come into existence – *if*, that is, that early fetus never in fact "become[s] a person."[178]

The upshot, according to Harman, is a "very liberal" position on early abortion: it requires "no moral justification whatsoever."[179] The "common liberal view," in contrast, often finds early abortion permissible but also holds that the early abortion "requires at least some justification, however minimal."[180]

The view that Harman describes exactly tracks the results on early abortion that I have proposed here. The early abortion (just like non-conception or conception followed by non-fruition) any ordinary case can be determined to be perfectly permissible and under a very summary analysis.[181]

4.5.3 Harman also wants to explain how the very fetus that has no moral status at a world where the early abortion takes place can also be an "appropriate object of love" at still another world, a world where the pregnant woman and her partner have no thought of or desire for an abortion.[182] Apparently accepting the idea that being an appropriate object of love and having some moral status come together, Harman uses the Actual Future Principle to unravel the puzzle. If there is no early abortion, and the fetus survives, it will develop into a person. And the "present nature" of that early fetus is determined by all the intrinsic properties it "ever has." Thus, the fetus that is not aborted – the fetus that later develops into a person – is a perfectly appropriate object of parental love now.[183]

4.5.4 Still, there are issues. Suppose that the woman in fact continues the pregnancy. One issue is how we are to evaluate her choice to do just that in the case where having the early abortion would not have made things worse for her or for anyone else. Harman wants to say that the early abortion itself is perfectly permissible. It "needs no justification." But if there is no abortion, then the early fetus itself has at least some "moral status." Moreover, Harman elsewhere argues, it is plausible to think that bringing a person into an existence worth having confers a

---

[178] Harman (2000), p. 311.

[179] Harman (2000), p. 313.

[180] Harman (2000), p. 313.

[181] Harman (2000), p. 313. Harman does leave open the possibility that the early abortion may require justification on grounds having nothing to do with the fetus. I assume that the reference here is to the odd case where the early abortion imposes a substantial loss on an individual other than the fetus. Harman (2000), p. 312.

[182] Harman (2000), p. 311.

[183] It is not clear to me why parental love would be appropriate in the case where there is no early abortion but never in the case where the early abortion does take place. After all, the woman may choose the early abortion very reluctantly – for the sake of, say, an already-existing child; she may in that case subjectively feel love for the fetus or even potentially for the merely possible person. And clearly, if parental love is appropriate in the case where there is no early abortion, then it is appropriate in the case where there is a miscarriage as well. Harman, however, considers and replies to objections along those lines. Harman (2000), pp. 319–320.

"benefit" on that person, and that we have a "moral reason" to confer that benefit by way of bringing that person into existence.[184]

The case, then, looks like this:

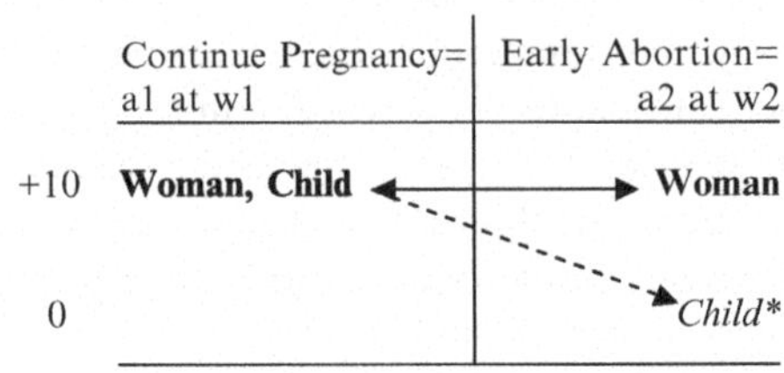

Case 4.1 The Actual Future Case

Harman's views about the moral status of the early fetus, the benefit of coming into existence and the role that moral reasons would seem naturally to play in our analysis seem clearly to support the result that a1 – continuing the pregnancy – is *obligatory*. But if a1 is *obligatory*, then we cannot say as well that a2 – the early abortion – is itself *permissible*. At least, according to standard deontic axioms, we can't have things both ways: we can't say a2 is *permissible* but that we are *obligated* to choose a1 instead.

4.5.5 On the face of things, Harman's view seems very like Exclusion Beta. In fact, Harman herself suggests such a connection, when she elsewhere describes the "Bad Method" as identifying – with Exclusion Alpha – the persons who *actually* exist and then concerning ourselves just with how our performed (actual) acts and unperformed (nonactual) acts affect those persons and those persons alone.[185] The Bad Method does seem bad. But what she calls the "Good Method" is not so great, either.[186] Applying the Good Method, we ask whether an act would harm any things. If it would, we then ask whether those harmed things have moral status.[187] And elsewhere: "one must be careful to evaluate actual events with respect to the actual moral statuses of the early fetuses involved; and one must evaluate *counterfactual* events with respect to the *counterfactual* moral statuses of the fetuses involved."[188]

On the face of things, the Good Method seems deficient in the same way that Exclusion Beta is deficient. The truth is that the evaluation of an act performed at

[184] See Harman (2004), p. 98 ("… we do have some reasons to create every happy child we could create. But there are reasons to benefit…. These reasons are very different from reasons against harming…."). See generally Harman (2009).

[185] Harman (2004), p. 106.

[186] Harman (2004), pp. 106–107.

[187] Harman (2004), p. 107.

[188] Harman (2000), p. 321.

one world is going to be strongly connected with evaluations of alternative acts performed at alternative worlds. It is that connection that deontic axioms often try to describe. Allowing the things that matter morally to shift from world to world, as Exclusion Beta explicitly does and as Harman's Actual Future Principle also seems to do, is going to generate an array of evaluations that just do not obey those axioms.[189]

4.5.6 In fact, however, it seems that Harman's view is more carefully drawn – and more complex – than this. On her view, that an agent has a moral reason to perform one act, no moral reason not to perform that act and no moral reason to perform any alternative act does not necessarily imply that the agent has an *obligation* to perform that one act. The fact that an act will create more wellbeing for some individuals who have moral status (e.g., the child at w1) and create less wellbeing for no one at all (e.g., the mother at w1) will not, on Harman's view, trigger the *obligation* to perform a1.

Harman's goal in introducing the notion of the contingent moral status of the early fetus is to show that a certain *necessary* condition for the early abortion's being wrong is failed. That is, for the early abortion to be wrong, the early fetus would need to have some moral status, and it doesn't. But none of this implies a *sufficient* condition on wrongdoing. Consistent with what Harman says, a sufficient condition on wrongdoing will be highly complex. It would go beyond the idea that, if we have a moral reason to do one thing, no moral reason not to do that thing and no moral reason to do anything else, then we are *obligated* to do that one thing. And it would go beyond the idea that if an act will create more wellbeing for some individuals who have moral status and create less wellbeing for no one at all, then that act is obligatory.

4.5.7 There are aspects of Harman's view, to be sure, that *suggest* that a1 is obligatory in the Actual Future Case. Harman thus writes that the choice not to abort means that the woman "*owes* [the early fetus] her love."[190] However, the relation between our having a *moral reason* to do something and our being *obligated* to do that thing is a complex one. The other side of the early abortion issue makes this very point.[191] Thus, the pregnant woman may have a good reason – a reason having to do with her own future wellbeing, a reason we consider a *moral* reason – to have an early abortion. Moreover, the early abortion may be perfectly permissible for her. But it is odd to think that she is *obligated* to have the early abortion. Surely, if she wants to have the baby, it is permissible for her to do so.[192]

4.5.8 There is good reason to think, then, that we should not attribute Exclusion Beta to Harman. At the same time, the more complex view that Harman seems to adopt is going make her position on early abortion itself more complex that it seemed starting out.

---

[189] See parts 2.2.1 and 2.6 above.

[190] Harman (2000), p. 317.

[191] I am grateful to Elizabeth Harman for this point.

[192] See, for example, Sider (1993), pp. 117–132.

It seems that, in Harman's view, *moral obligation* in any strict sense is going to be hard to come by. If a1 is not obligatory in the Actual Future Case, notwithstanding all there is to be said in favor of a1 – the moral reasons that favor a1, the moral status of the child who is benefited by a1 – then the test that Harman sets for obligation is a stringent one. Little is obligatory. But if that is so, then much is *permissible*, including a2.

But that means that a2 is permissible even though the agent has a moral reason *not* to perform a2. At least, it seems that the bare fact that the agent has a moral reason to perform a1 means that the agent has a moral reason *not* to perform a2. And we have already said that the agent does have a moral reason to perform a1.

But that means that when we counsel the young woman that the early abortion is *perfectly permissible* – that it requires "no moral justification" – the young woman must now turn to the task of reconciling the permissibility result that *allows* the early abortion and the moral reason *against* the early abortion. The complexity of the assessment is itself going to make her, and us, wonder about the very meanings of the terms that are being used. Can something really need "no moral justification" when we have a moral reason *not* to do that thing?

4.5.9 The opposing theses in Harman's work aren't inconsistencies. But they do suggest an *argument of sorts* against early abortion.

Consider, again, the Actual Future Case. Harman wants her view to suggest that a1 and a2 are both permissible. So far, so good. But her view also suggests that the agent has a moral reason to perform a1 and hence a moral reason not to perform a2, and no moral reason not to perform a1. If this is moral law – and Harman doesn't claim that it is, but if it is – then it is not clear what it is telling us. Something like: both alternatives are permissible, but the one is really *better* than the other. Or: neither choice is *really* terrible, but a2 is a little bad. Or: the agent who aims to act as moral reason instructs will choose a1 and leave a2 alone.

4.5.10 Is there a way to make things easier? Or must we instead just live with the idea that moral law is complex – and that the morally cautious pregnant woman would do well to avoid the early abortion in any ordinary case?

Variabilism suggests one strategy. The intuition behind Harman's claim that early abortion "requires no justification" seems to be that at any world where the early abortion is in fact performed that choice imposes no harm – no loss – on any existing or future individual that has any "moral status" at all. This intuition is – in part – shared by Variabilism. The zero wellbeing level implied by never having been brought into existence to begin with – that *loss* – just isn't morally on the same plane as the loss that is imposed on the person who exists and is treated so badly that his or her overall wellbeing level is reduced to zero. The Basic Case makes that point. If a person incurs *any loss at all* in virtue of never having been brought into existence, then that loss, according to Variabilism, has *no moral significance whatsoever.*

So far, the views coincide. But now let's turn to the benefit side of things. For Harman, the agent – at, for example, w1 in the Actual Future Case – has a moral reason to confer the benefit of an existence worth having on the child. This is to say that that particular benefit – that gain for the child – itself has moral significance.

Now, it might seem difficult to say that that gain does not have moral significance, since the child in fact *exists* at w2. Variabilism, however, in fact has the resources to say just that. For according to Variabilism the child matters morally, but matters *variably*. That means that just as *some* of that child's *losses* have moral significance and *some* do not, *some* of that child's *gains* have moral significance and *some* do not. After all, losses and gains – intuitively, and as a metaphysical matter – mirror each other. If one act creates a *gain* relative to an alternative, the alternative creates a *loss* relative to the one.

Harman's view seems to be that a *gain* for an individual that itself has moral status can create a moral reason to perform an act, but that the corresponding loss does *not* create a moral reason *not* to perform the alternative act. In contrast, according to Variabilism, we can take the position that, just as the *loss* the child incurs in w2 is devoid of moral significance, so is the *gain* the child enjoys in w1 devoid of moral significance.

Variabilism thus denies that the gain – the benefit – that is created for the child at w1 has any moral significance at all. We can, accordingly, deny that there is any moral reason for the woman to choose a1 – or not to choose a2. We thus avoid the need to explain why a2 is permissible even though the pergnant woman has a moral reason not to choose a2. We simply deny that she has any moral reason not to choose a2.

4.6 Distinction Between Variabilism and Its Competitors

4.6.1 Each of the views described above sets aside the issue of just when the new person comes into existence – just when, that is, we are dealing with a *person* rather than simply a live human organism.

Hare relies on the Golden Rule – but notes that the results he reaches regarding abortion conform to the "classical utility principle," that is, Totalism. Under both those approaches, the merely possible person matters morally. And that person matters morally, not just *variably*, but rather *necessarily*: the loss that person incurs, when we fail to bring that person into existence, has every bit as much moral significance as the loss that same person incurs at a world where he or she exists.

Marquis, too, considers the merely possible person to matter morally. On Marquis's view, however, the merely possible person matters morally, but matters *contingently*: the loss of the future of value – the future that the merely possible person would otherwise have enjoyed – has moral significance *only if* the *individual* whose future of value is at stake in fact exists.

Harman, too, relies on the contingency of existence – but she focuses on the contingency of the *future person's* existence, not the contingency of the *individual's* existence. Accordingly, Harman, in contrast to Marquis, makes the *merely* possible person – the person who in fact does not and will never exist – matter morally not at all. For her, it is the fact that a person *will* exist that makes that person matter morally

– that the person is a *future* person, and not a *merely* possible person – that imbues any losses that person incurs with moral significance.

On none of these views, then, is it critical *when* the person comes into existence. Each view focuses, instead, on whether that person *could have existed* or *will exist*. Intuitively, though, timing seems central to the question of abortion (and infanticide and indeed in many cases euthanasia): just *when is it* that what we are choosing to destroy is not simply a live human organism but rather a thing that moral law itself requires us to create more wellbeing for rather than less.

Thus, it is at least arguable that we can create more wellbeing rather than less for the early human fetus. After all, in such a case we do have a live human body, something that may continue to develop and flourish depending in large part on how we now conduct ourselves.

But we should resist the idea that we are obligated to create additional wellbeing for anything we *can* create additional wellbeing for. The fact that we can, in a way, create additional wellbeing for the early fetus does not imply that we ought.

Hare, Marquis and Harman do not deny, of course, that at some moment in time – or perhaps over some period of time – a person emerges during the pregnancy or at some later stage, during infancy or childhood. What they deny is the moral significance of the timing of that event.

4.6.2 Variabilism, in contrast, spotlights the timing of the person's coming into existence in relation to the abortion's being performed. It isn't the existence of the person per se that is important, of course: such a view would seem tantamount to Exclusion in either its Alpha or Beta form. According to Variabilism, what is important, instead, is whether the loss that the person incurs is incurred at a world where that person does or will exist. If the abortion takes place at a world *prior* to the moment at which the person comes into existence – if it keeps that person from ever existing at all – then any loss that person incurs at that world is devoid of moral significance since that person will never exist at that world. In contrast, if the abortion takes place at a world *after* the moment at which the person has come into existence, then any loss incurred by that person will have full moral significance since that person does or will exist at that world. Any such loss will then count against the act that imposes it.

A respect in which this approach seems intuitively plausible is that it focuses our attention on what is going on with the fetus in utero. Is it yet capable of feeling pain? Or is its neurology simply developed enough to show a mechanical reaction to a pain-stimulus? Is it thinking, rudimentarily or even sub-consciously? Is there a psychological, or cognitive, being there, something that can be pulled one way or another by the activities and responses of its own mind, that can feel aversion to certain events and pleasure in others?

We think – I think we think – that it's wrong to torture a cat for no reason whatsoever and that it's wrong to torture a human baby for no reason whatsoever. That means that those beings have moral status. In deciding whether to torture or not, we must take their interests into account alongside our own. The question that we intuitively want to focus on in thinking about abortion, and the question Variabilism brings to the fore, is just when it is that the developing human organism acquires that same moral status – just when it is that a *person* comes into existence. Variabilism thus considers morally relevant what we intuitively think *is* morally relevant.

Chapter 5
Abortion and Variabilism

5.1 Introduction

In what follows, I argue for three claims. First, what I will call *early abortion* is ordinarily, and almost always in the case where it is what the pregnant woman wants, permissible. And that is so, whether she wants the early abortion for a good reason, a poor reason or no reason at all.

Second, what I will call *late abortion* is ordinarily wrong. That is so, even in cases in which the pregnant woman has a very good reason for wanting to have the late abortion.

My third claim is that the first two claims themselves represent an attractive *middle ground* on abortion – a position that takes what is most clearly plausible from the more extreme positions on either side of the abortion debate and leaves the rest aside.

The account of abortion that I propose pivots on the timing of the abortion in relation to the timing of a new person's coming into existence. Accordingly, I begin by considering what persons are and what signals their coming into existence.

5.2 What Is a Person?

5.2.1 For purposes here, to be a *person* is to be the *kind* of thing we have obligations in respect of – the *kind* of thing we often must make things better for – *at least* in the case where that thing exists. I also accept, for purposes here, that persons are *thinking things, broadly construed.* Perhaps the reason it seems so plausible that *thinking* and *mattering morally* come together is just that it *matters to thinking things* (whether they think so or not) that we treat them in one way rather than another. It is hard to see how it can matter to the thing that never thinks how that thing is ever treated.

5.2.2 In this context, just how broadly is *thinking* itself to be construed? While mechanical reactions to pain stimuli will not count as thinking, any *felt* pain will. So will the thinking that we do that takes place beyond our own immediate awareness. We thus need to take into account the increasingly plausible notion that *we* are unaware of much of the cognitive work that *we* do. We *grasp* things – we note them, take them into account, figure them out – before we realize what we have done. And there would seem to be some unitary self there doing the work or at least the

M.A. Roberts, *Abortion and the Moral Significance of Merely Possible Persons: Finding Middle Ground in Hard Cases*, Philosophy and Medicine 107,
DOI 10.1007/978-90-481-3792-3_5, © Springer Science+Business Media B.V. 2010

recording of the work. Otherwise, the discrete stages of a sequence of thoughts that we may at some point consider to have ended in an insightful or valuable way of looking at things would remain just that: discrete stages incapable of propelling anything of interest into conscious awareness.[193]

These points underline what (for all we know) is a metaphysical necessity: that the felt pain requires a subject that feels; that thinking requires a thing that thinks. For purposes here, I will understand the term *person* to include not subjectless thinking, if there can even be such a thing, but rather just thinking subjects. But I do not assume that thinking subjects necessarily have a concept of themselves as ongoing subjects of conscious experience – or that they have any concept of *themselves* at all. It's not my *concept* of myself as an ongoing subject of conscious experience that makes me a thinking thing and the kind of thing that matters morally. What is critical is that *there exists* such a subject.

5.2.3 Must the thing that thinks have acquired the capacity to *survive* – to continue in existence as the *same* thinking thing – in order to count as a *person*? Must it be a sufficiently hardy cognitive being that its own thoughts at one moment can bind together with later thoughts in the way that is, in the case of persons, required for their survival? Must that thing be such that the relations of *psychological continuity* that seem critical for personal survival and identity across time will – depending on what *we* do – obtain going forward?

This is a hard question. But it is not one we need to answer here. For we should not consider the act that removes a person from existence when that person was *bound by nature to go out of existence* in any event – bound, that is, to go out of existence *regardless* of the conduct of any agent or any group of agents – to have imposed a *loss*. Thus, if the *present* subject of the felt pain is just too fragile to sustain the required psychological continuity with *any* future subject of a felt pain – if the present subject will inevitably, naturally fade from existence when the pain subsides – then that subject will not incur a loss when it is removed from existence by the early abortion. That is so, even if that subject is itself a thinking thing and a person. It is on its way out of existence *no matter what anyone does*. There is no world accessible to agents in which that subject continues to exist. In such a case, the early abortion ends the existence of a person – a person whose existence was bound to end in any event – without imposing a loss on that person.

Now, there is a caveat: there is nothing in what we have just said that implies that it is permissible to perform an early abortion *in a way that itself imposes pain*. That is, even the person who has not yet acquired the capacity to survive may still have the capacity to feel pain. And any pain that the early abortion imposes on the non-continuing person will count as a loss – and, under Variabilism, as a morally significant loss that will need to be weighed against other morally significant losses.

---

[193] "[I]dentity is nothing really belonging to these different perceptions, and uniting them together … ." Hume (1975/1739–40), p. 260 and, generally, pp. 251–263; see also pp. 633–636 (questioning even the uniting, or bundling). "At a certain point, you just have to admit that your brain knows much more than you do." Jonah Lehrer (2008), pp. 40–45 (quoting Mark Jung-Beeman, and citing Poincaré's own description of the processes that led to his particular insight into non-Euclidean geometry, which happened, Lehrer notes, while Poincaré was "boarding a bus").

Still, the upshot is that the only persons – the only thinking things – we shall need to concern ourselves with here for purposes of evaluating the early abortion *insofar as that early abortion eliminates a person from existence* are those persons who have *acquired the capacity to survive from one moment to the next* – those persons who, depending on what *we* do, will survive from one moment to the next.

Just to put these points in perspective, it may well be that the stage at which we have a person but a person who has not yet acquired the capacity for survival is *very* brief. It may well be that, soon after a thing begins to think – in the sense required here; not in the sense of displaying electrical activity in the brain, or reacting to pain stimuli – it acquires the capacity to survive.

5.2.4 The view that I am adopting here of what it is to be a *person* is an inclusive one. My aim is to capture a sense of the term *person* that includes all and only those things in respect of whom it *matters morally* how we treat them. This concept of person is clearly not limited to adult human beings who are engaged in sophisticated thought processes. It may include what Mill described as the "whole sentient creation."[194] Many non-human animals will count as persons, and many human beings and much that might be described as "early human life" will not.[195]

Photo 1 *A person?* A four-and-a-half week-old (est.) orphan kitten

[194] See Chapter 2 of John Stuart Mill, *Utilitarianism* ("What Utilitarianism Is"). Mill 2002/1863.

[195] While it is not a concept that meshes well with Peter Singer's own concept of what counts as a person, it does recognize that he is surely right to think that our moral obligations extend to many non-human animals – but not to each and every thing that counts as human. Singer (1999), pp. 83–95.

Photo 2 *A non-person*? "A 10 mm (0.39 in.) embryo from an ectopic pregnancy, still in the oviduct. This embryo is about 5 weeks old (or from the seventh week of pregnancy)." From "Embryo," *Wikipedia* (retrieved March 15, 2010)

On this view, rats, which think, are persons. They matter morally. It does not, however, follow that the obligations we have in respect of each and every one of the world's rats are going to be identical to – or as onerous as – the obligations we have in respect of all of humanity. That is so, to some extent, simply because rats are in their own way less demanding than cognitively active human beings are. Their interests are, plausibly, less difficult to protect.

But another part of the puzzle is that we do not have the hope for them, or the expectation, that we have for human beings. Where we must choose between saving the human being who will save the lives of thousands more human beings, and the human being who will do just the reverse, we must choose the former. Where we must choose between saving the dog who will save the lives of thousands of human beings, and the human who will do just the reverse, we plausibly must also choose the former.

Still, even if our duties in respect of rats turn out to be less onerous than our duties in respect of cognitively active human beings, the claim that rats are persons (in the moral sense I am aiming for here) has a clear practical implications. It means that we are obligated to conduct ourselves in a certain way in respect of them – that their interests, alongside our own, must be taken into account in determining what we ought to do and that treating them badly can ground a finding that what we have done is wrong.

5.3 When Do Persons Come into Existence?

5.3.1 In determining when a new person comes into existence, it's useful to consider just when we think that *we ourselves* came into existence and what it will take for us to go out of existence. It seems plausible that each of us commenced existence at the moment of our first thought and each of us will stop existing at the conclusion of our last. More generally, it seems plausible that persons come into existence at that moment at which they first think and go out of existence at that moment at which their lost thought concludes.

5.3.2 To take into account the fact that the future can unfold in different ways and that there may be a time at one world where a particular person first thinks but no such time at another world, we can put this point as follows:

> A person p comes into existence at a world w at the moment at which p at w begins to think and goes out of existence at w at the moment that p's last thought at w is itself concluded.

Now, there is nothing in this view that requires that persons must think *continuously* in order to exist. Consistent with this view, it is enough for them to think *intermittently*.

When people are in a coma or undergoing general anesthesia, perhaps they do not think at all, even subconsciously. That a thing *continue to think* is not plausibly a necessary condition for a person to *continue to exist*, that is, to *survive*, from one moment to the next. What thinking does instead is define a particular person's beginning and ending points. Persons begin to exist with their first thought and end at the conclusion of their last.

5.3.3 On this view, a thing that exists at a world w but never has a first thought at w by definition is not a person. A thing – for example, the human gamete, or the human embryo – that is soon to be washed from the womb, either via an abortion or some natural process, is not a person because it never has that first thought. Nor is a thing – for example, a corpse – that has already had its last thought – if indeed we want to say that the *body*, apart from the *person*, ever thinks at all – is not a person. If it continues to exist past "its" last thought, then it is not a person.

Thus, the embryo comes into existence soon after conception. But the person does not come into existence until later. The corpse may linger when the person is gone. The principle of the indiscernibility of identicals entails that the embryo and the person, and the corpse and the person, are distinct individuals.

5.3.4 If the live, healthy, human body isn't a person, can it ever think? I think the view that it can't would be hard to defend, given that the two seem to be – for a limited time – perfectly overlapping in space. So, when we say a person is a thinking thing, we are saying, not that thinking is *sufficient* for a thing to be a person, but rather that it's *necessary*.

But thinking *is* sufficient to show that there is a person *in the vicinity* (though, if that thing has not yet acquired the capacity to survive, the relevance of that fact for purposes of evaluating the early abortion will be limited). If a particular human body is thinking – reading the newspaper, writing a poem – that seems enough to tell us that, at least for those moments, that body is intimately correlated with a particular person with whom it overlaps in space.

5.3.5 At the same time, the fact that a live human body – for example the human fetus – will if "left alone" think *at some later time* is not sufficient to make that live human body a *person* – or to show that there is, at the *earlier* time, a person *in the vicinity*. In other words, that a thing *x* will if "left alone" at a *later* time overlap for awhile, spatially, with a thing *p* that is a *person* – that has had that first thought – does not mean that *x* is a person *or* that *p* itself has, as of the *earlier* time, come into existence.[196] Similarly, that a thing *x*, when alive, may overlap spatially for awhile with a thing *p* that is a person – but whose last thought has already taken place – does not mean that *x* is a person *or* that *p* continues to exist. The human body *x* may survive death – assuming that the body is not annihilated along with the person, as in an explosion – but the person *p* will not.

Thus, the body that constitutes a particular human corpse at a later time can plausibly be said to be identical to the spatiotemporally continuous human body that constituted a live human being moments before. And, plausibly, that *body* was itself cognitively active at the earlier time. But the corpse is not identical to the person. The corpse is here, the person is gone.

Similarly, the fact that the early fetus – a continuing human organism – engages in thinking *at some time or another* does not entail that *the fetus itself* is a person. The fetus 18 weeks into the pregnancy may well be identical to the fetus 30 weeks into the pregnancy. But the fact that the fetus thinks at *30* weeks – assuming that it does – doesn't mean it is a *person* at *18* weeks. Nor does the fact that the fetus thinks at 30 weeks mean that the *fetus* is a person at *30* weeks. Rather, it just means that at 30 weeks (but not 18) there is a person *in the vicinity* of the fetus: the fetus simply overlaps spatially for awhile with the very person who eventually "develops from" the fetus at 18 weeks.

5.3.6 As the foregoing discussion suggests, that a particular human embryo has been created, and will if "left alone" *develop into* a particular person, does not imply that that *person* has come into existence. There has been no first thought. If the pregnancy is then terminated, naturally or otherwise, then the person who *might* have existed – the *merely possible person* – will never in fact exist at all.

5.3.7 By definition for purposes here, the *early fetus* is limited to the fetus that has never had that first thought that signals the coming into existence of a *person*. And early abortion is simply the abortion of the early fetus, or indeed the embryo, or the abortion of a person who has not yet acquired the capacity to survive as the same thinking thing.

The question of *just when that first thought happens* is critical to any practical application of Variabilism in the context of abortion. But that is not a question I am going to fully explore or try to resolve in this book. For purposes here, I am instead going to describe some of the basic elements of a plausible timeline. Of course we want to know more; among other things, we want a clearer idea of when that first thought itself is likely to occur. But some things we know now.

The physical basis for thinking seems not to be in place until at the earliest the 20th week after conception. At least, the underlying structures necessary for the felt pain

[196] See Lynn Rudder Baker (2007).

are not in place until still later – around the 29th or 30th week after conception.[197] These points suggest that there is a substantial period of time between conception and the moment of that first thought during which the termination of the pregnancy involves the destruction of an organism that is genetically human and is alive and perhaps thriving but that is not – in contrast, perhaps, to the 5-week-old kitten – a *person*.

5.3.8 For purposes here, the *late fetus*, by definition, is limited to the fetus that has already engaged in that first thought and that has already acquired the capacity to survive. The late fetus is, in other words, something we can correlate not just with a *possible* person but rather with an *existing* person who will survive if left alone. Late abortion, then, by definition is just the abortion of the late fetus.

Again, we face the question of just when that first thought takes place. Working backwards, I make the assumption that intact, full-term human infants are thinking things, broadly construed. They seem clearly to feel pain and pleasure. They have memories and expectations, of sorts, and may be engaged in unconscious work as well.

It is true that there has been some debate on whether newborns have the capacity to feel pain. More recently, however, the sense has been that they do.[198] Thus, we may decide not to give the newborn an anesthetic to blunt the pain of circumcision since many anesthetics come with risks of their own. But the infant will still cry when circumcised, and we know its neurological apparatus looks much like ours. There is no reason to think the experience of pain – the *felt* pain; not what we see when the cerebral cortex has been destroyed but the brain stem remains intact, making it possible (as in the case of Terri Schiavo) for the eyes to open and the body to flinch – exists for us but not for the newborn.[199]

---

[197] The question of timing has been investigated for the case of the felt pain; it is possible, of course, that thinking itself might precede that point. On the question of the felt pain, the following note is pertinent: "[T]he capacity for conscious perception of pain can arise only after thalamocortical pathways begin to function, which may occur in the third trimester around 29–30 weeks' gestational age, based on the limited data available. Small-scale histological studies of human fetuses have found that thalamocortical fibers begin to form between 23 and 30 weeks' gestational age, but these studies did not specifically examine thalamocortical pathways active in pain perception. While the presence of thalamocortical fibers is necessary for pain perception, their mere presence is insufficient – this pathway must also be functional. It has been proposed that transient, functional thalamocortical circuits may form via subplate neurons around midgestation, but no human study has demonstrated this early functionality. Instead, constant [somatosensitive evoked potentials] appear at 29 weeks' [postconceptual age], and EEG patterns denoting wakefulness appear around 30 weeks' [postconceptual age]." See Susan J. Lee et al. (2005), p. 952.

[198] See, among many other sources, Taddio (1997), pp. 599–603.

[199] We need to draw a distinction between the "persistent vegetative state" – the phrase used to characterize the body of Ms. Schiavo – and the "minimally conscious state," in which the individual's MRI-read cerebral cortex displays the same sorts of reactions to, for example, childhood stories read aloud to them that our own perfectly awake brains would display. "Like an interlocking set of old Christmas lights, blinking on and then off, the neural connections in minimally conscious patients seem to be in place, the research suggests. In persistently vegetative brains, by contrast, the crucial connections are apparently shot: maybe one light blinks here, another over there, but the full network is dark." Benedict Carey (2005).

It is, moreover, implausible that human infants become cognitively active exactly at the moment of birth. While this logic cannot be applied indefinitely, it does seem clear that if human infants think just after they are born, they surely think just prior to birth as well. Surely they feel pain and pleasure and have memories and expectations, of sorts. And they are perhaps engaged in unconscious work – making their way, consciously or not, through some basic problems of space and time (no doubt particularly space) they encounter in utero.[200] The fetus that is just a few days or weeks away from delivery seems clearly to fall into the category of thinking things.

5.3.9 For purposes here, I will use the term *middle fetus* – and *middle abortion* – to capture just those cases where we really do not know whether that first thought has taken place or, if it has, whether the individual has acquired the capacity to survive. Here, we do not know whether the abortion itself is the choice not to bring a new person into existence or is instead the choice to end the life of an existing person.

I am taking it for granted here that that first thought happens at a determinate time. But it may never be quite certain for us just what that time is. We may know that it does not happen before, say, the 20th week after conception. But we may not know that it does not happen before, say, the 30th.

The middle abortion thus represents an epistemic category, not a moral one. For purposes of determining the permissibility of the middle abortion, we do not need a more complex moral analysis but rather need to know more about fetal development.

5.3.10 Now, at one level, the same is so for the early and late abortion as well. We need to recognize the middle abortion; we need to recognize that there is much that we do not know about fetal development. And we should recognize that that is so both earlier and later in the process of fetal development as well.

As I am using that word here, minimally conscious individuals may constitute "persons." It just isn't clear. That will not in itself determine the issue of whether they must receive the hundreds of thousands of dollars worth of medical care that would be necessary to give them a chance at full consciousness or even keep them alive. In contrast to the scenario in which person never comes into existence at all as a result of the early abortion, we here face a *tradeoff* scenario. Suppose that we can, at very great cost, do a little bit of good for the person who is in the minimally conscious state. We can give that person a chance of recovery and, at very great cost, we can in the meantime care for that person's bodily needs and accord to that person whatever pleasure we think the sedated minds might be capable of. But suppose that we can do that much for the one person only by depriving many other persons of badly needed health care, nutrition and housing, forcing upon them wellbeing levels at or below that of the one person. It is not at all clear that, having established that minimally conscious individual is in fact a person means that we must do the *best that we can* for that person, whatever the cost to others. In fact, it seems that just the reverse is true. We will come back to such tradeoff scenarios in what follows. For purposes here, the important point is just the fact that a thing is a person and, at a world *w*, does or will exist, means just that that person's interests must be taken into account – that any loss he or she incurs at that world is morally significant and *bears on* the evaluation of that act and its alternatives. The fact that the minimally conscious individual is a person – and an existing person, at that – means that there is a morally significant *tradeoff* in the offing that will need to be made. It does not, in itself, tell us how to make that traeoff.

[200] Lehrer (2008), p. 43. See note 193 above.

The picture is not one that I will try to complete for purposes here. On the other hand, it isn't useful to adopt an air of skepticism on the question of whether there *really is* such a thing as early abortion – the abortion, that is, of an embryo or fetus that has never yet had that first thought (or, if it has, has not yet acquired the capacity to survive). It isn't useful for us to take the position that, for all we know, the newly formed human embryo is a thinking thing – or that the newly formed fetus, for all we know, can feel pain despite the fact that its thalamocortical pathways aren't yet in place.[201] We cannot pretend to know *nothing* about fetal development. Nor is it useful to be overly skeptical on the question of whether there really is such a thing as late abortion – the abortion of a fetus that has had that first thought and that has acquired the capacity to survive.

Thus, we know enough to be sure that the human embryo and the human fetus in the early weeks of pregnancy are not the kinds of things that think – not, that is, *persons*. But we also know enough to be sure that the relatively healthy, full-term or nearly full-term fetus very likely is exactly that kind of thing – a thinking thing, a *person*, that has acquired the capacity to survive as the *same* thinking thing.

5.4 Early Abortion: Three Cases

5.4.1 *Early abortion* is the abortion of a human embryo or early human fetus that, while indisputably human and alive, is not yet a person or at least has not yet acquired the capacity to survive from one moment to another.

My claim about early abortion is that it is often, and almost always in the case where it is what the woman wants, perfectly permissible, whether the woman chooses it for a good reason, a bad reason or no reason at all. Three types of cases warrant consideration.

5.4.2 *Early abortion/Case I* supposes that no one other than the person the early fetus is likely to develop into if the early abortion is not performed has any stake in how the choice is made. In other words, suppose that the wellbeing levels of the pregnant woman, her partner, her *grandmother* and each other person are unaffected by how the choice of early abortion is itself made.

The early fetus is not itself a person. But it can be *correlated* with a particular person – a person who may well eventually exist at a world where the early abortion is not performed and will never exist at a world where the early abortion is performed.[202] The early abortion thus does not end the life of an already-existing person but rather means that the person the aborted fetus is correlated with will never come into existence to begin with. At the world at which the early abortion is performed, in other words, the early abortion does nothing more than relegate that particular person to the status of the *merely* possible relative to that world.

---

[201] See note 197 above.

[202] The newly conceived embryo that has a fission in its future can, similarly, be correlated with two or more persons – all of those persons who will in fact come into existence as a result of twinning.

Now, the account of *loss* I have adopted for purposes here means that the merely possible person typically incurs a loss at the world where the early abortion is in fact performed.[203] But that we recognize the person typically incurs a loss does not compel us, according to Variabilism, to recognize that that loss has any moral significance whatsoever. Thus:

Variabilism:

A loss incurred at any world w by any person p has moral significance for purposes of determining the permissibility of any act a performed at w that imposes that loss and any alternative act a' performed at any accessible world w' that avoids that loss *if and only if* p does or will exist at w.

By implication, the loss incurred by p at w when agents fail to bring p into an existence worth having at w will have no moral significance for purposes of determining the permissibility of the act a performed at w that imposes that loss or any alternative act a' performed at w' that avoids that loss.

The loss incurred at the world where the early abortion is performed is, accordingly, devoid of moral significance for purposes of evaluating the early abortion or any of its alternatives. That is so, according to Variabilism, since the person who incurs that loss never exists at all that world.

The otherwise plausible permissibility theory, constrained by Variabilism, is thus barred from counting that loss *against* the early abortion. Having no moral significance, that loss cannot, on any *plausible* permissibility theory, make the otherwise permissible choice of abortion *wrong*. By assumption, no other losses – at least no other morally significant losses are at stake; not the woman's, not her partner's, etc. The otherwise plausible permissibility theory accordingly declares the choice of early abortion permissible.

Specifically, OPPP2 is useful in this context.

Otherwise Plausible Permissibility Principle 2 (OPPP2):

An act a performed at a world w is wrong *only if* act a imposes a morally significant loss on a person p at w.

The woman can, on a whim, have the abortion. She can have the abortion because she knows she will not otherwise be able to fit into an especially nice dress a couple of months hence. She can have the abortion because she plans to leave in a few weeks for study in France. She can have the abortion because she doesn't feel like telling her boyfriend that she is pregnant. She can have the abortion because 2 + 2 = 4. She can have the abortion because the coin she has flipped has come up heads. She can have the abortion for any reason or for no reason at all. The woman need not establish that not having the early abortion will cause her or another to incur a morally significant loss, since there is no morally significant loss on the other side of the scale that she needs to balance things against.

---

[203] See parts 2.2.1–2.2.3 above.

It is a plus for Variabilism that it can consider the loss the merely possible person incurs at a world where that person never exists at all not to have any moral significance, without taking the more extreme view that *none* of that person's losses ever have any moral significance. At worlds where the early abortion is not performed – and where the person eventually does exist – any losses that person happens to incur will have full moral significance for purposes of evaluating the act that imposes that loss and each of that act's alternatives. Variabilism thus does not make the mistake of radically *excluding* the merely possible from the realm of those who matter morally.

This means, among other things, that acts that impose losses at a world *w′* on the person *p* who, as a result of the early abortion, never exists at all at the one world *w* may well be wrong. Thus, suppose that *p* does or will exist at w′. And suppose that the act under scrutiny at *w′* is the act of *continuing the pregnancy*. And suppose as well that the person has a life that is *less* than worth living – that is, a genuinely *wrongful* life – at w′. Then, according to the account of loss I have adopted for purposes here, the comparison between *w* and *w′* means that *p* incurs a loss at w′. Moreover, according to Variabilism, that loss has full moral significance. It bears on the permissibility of the act performed at *w′* that imposes that loss – and the permissibility of each alternative to that act that avoids that loss. It counts *against* what is done at w′ – that is, continuing the pregnancy – and, in a roundabout way, *in favor* the early abortion.

On the assumption that no other morally significant losses are at stake, OPPP1 – which could be aptly named *Pareto Minus* – implies that what is done at *w′* is *wrong*.

Otherwise Plausible Permissibility Principle 1 (*OPPP1*):
An act *a* performed at a world *w* is wrong *if*

(i) Act *a* imposes a morally significant loss on a person *p* at *w*;
(ii) An act *a′* at an accessible world *w′* performed in lieu of *a* at *w* is better for *p* at *w′* than *a* is for *p* at *w*; and
(iii) Either (a) *a′* imposes no morally significant loss on any person *q* at *w′* other than perhaps *p*, or (b) if *a′* imposes a morally significant loss on a person *q* at *w′*, then *a* imposes a morally significant loss on *q* at *w*, and *a′* is at least as good for *q* at *w′* as *a* is for *q* at *w*.

Deontic axioms then tell us, as well, that what is done at *w* – the early abortion – is itself not just permissible but also *obligatory*.

The upshot is that, while the woman may permissibly have the early abortion even though she has no good reason, or even no reason at all, not to have the baby, it does not follow that she may permissibly have the baby, for no good reason or for no reason at all. That Variabilism can deal with the case of wrongful life is important. At the world at which the woman makes the choice to continue a pregnancy when she knows that the baby, as a result of a disorder that we cannot treat, will unavoidably have a life less than worth living, the baby – that *person* – both exists and suffers. Variabilism will declare the loss the baby incurs at that world to have full moral significance. And barring exceptional circumstances, the otherwise plausible permissibility theory – specifically, OPPP1 – will imply that the choice to continue the pregnancy at that world is wrong.

But we do not need the case of wrongful life in order for Variabilism in combination with the otherwise plausible permissibility theory to generate a finding of wrongdoing. In the case of the genuinely wrongful life, the outcomes described by w and w' exhaust the agent's alternatives. The woman can either have the early abortion or continue the pregnancy and cause her child to incur a very great and very deep loss. There is no third option w'' where that child both exists *and* has a life worth living. But in many other cases there will be a third option, a way of bringing the child into a much better existence. As in Addition Plus, the combination of Variabilism and the otherwise plausible permissibility theory – again, OPPP1 – will ground a finding of wrongdoing.

5.4.3 *Early abortion/Case II*. Of course, the cases can get more complicated. Let's suppose that another person – beyond the person we correlate with the early fetus – does have interests at stake in how the choice is made. Suppose, for example, that the pregnant woman herself will incur a loss if the abortion is not performed. Her loss will be deemed morally significant, according to Variabilism, since she obviously exists at the world where that particular loss would be incurred. Suppose, as well, that no one, other than the person whose coming into existence is at stake, will incur any loss whatsoever if the early abortion is performed. (Suppose, in other words that the early abortion maximizes wellbeing for each person who does or will exist, at the world where the early abortion is performed.)

We now face a *conflict, of sorts*: more wellbeing can be created for the pregnant woman *only if* less wellbeing is created for the person we have correlated with the fetus. But the loss the pregnant woman faces, according to Variabilism, has *full moral significance*, while the loss the merely possible person incurs at a world where that person remains merely possible – where, that is, the early abortion is performed – has *no moral significance whatsoever*. The otherwise plausible permissibility theory easily prescribes the tradeoff that must be made. The early abortion remains permissible, under OPPP2. But continuing the pregnancy is now impermissible, since there is a way of making things better for the pregnant woman, and avoiding the morally significant loss on her behalf, without imposing a morally significant loss on anyone at all. OPPP1 implies that not having the early abortion is wrong. Deontic axioms tell us more: the early abortion is *obligatory* since the woman is left here with only one permissible choice.[204]

---

[204] Some theorists suggest that it "obviously" permissible for the woman not to choose abortion against her own interests. McMahan (2006), p. 636. See notes 121 and 126 above. But that is not so clear to me. For one thing, this is easily a confusing case, since no one plausibly thinks that it would be obligatory for a surgeon to perform an early abortion on a woman who has chosen to continue the pregnancy. So to avoid that sort of issue, let's suppose that the pregnant woman is undergoing surgery unrelated to her pregnancy for which an anesthetic is required, and that prior to the procedure she has said (correctly) that she is neutral on whether the pregnancy is to be continued or not, and that that is so even if an emergency arises and continuing the pregnancy – not having the early abortion – will mean that she will incur some significant loss. According to the view now under inspection, it is permissible for the woman to make the choice to continue the pregnancy. Now let's evaluate the choice whether to allow the woman to continue the pregnancy from the perspective of the surgeon. An emergency arises, and the surgeon must now decide whether to perform the early abortion, where the early abortion is now essential to the woman's continued

5.4.4 *Early abortion/Case III.* It can also happen that it is in the interest of the pregnant woman to have the early abortion and in the interest of another person who does or will exist – her partner; her grandmother – that she not. The reverse can happen just as well. But let's start with the case where the pregnant woman will incur the morally significant loss if the early abortion is not performed, and another person – her partner, her grandmother – will incur a morally significant loss if the early abortion is performed.

The first question is whether the case really does involve a conflict. By hypothesis, the woman incurs a loss at – say – w, and the other person incurs a loss at – say – w'. If there is some third world w'' accessible to the agents that avoids the conflict altogether, without imposing any morally significant loss on anyone at all, then there is no genuine conflict. If, in other words, the loss that the other person is at risk of incurring can be avoided by some means short of the woman's continuing the pregnancy, then the case does not involve a genuine conflict. Agents can just do what it takes to bring about w'' – that third outcome – instead.

But suppose there is a genuine conflict. Then, we will need a principle that goes beyond OPPP1, since OPPP1 is limited to the case where agents can make one person – for example, the pregnant woman – better off without imposing a morally significant loss on anyone at all. We also need a principle that goes beyond OPPP2, and addresses not just *conflicts* but *morally significant* conflicts – a principle that

---

good health. It seems that performing the early abortion is obligatory for him. Now, there is no contradiction here; we aren't saying that one act is both permissible and wrong. For the one act – continuing the pregnancy, by way of signing an appropriate consent form prior to the procedure – is the woman's, while continuing the pregnancy, by way of not performing the early abortion once the emergency arises – is the surgeon's. Still there is an awkwardness to the view that continuing the pregnancy can have one moral status relative to one agent, and another moral status for another. It means moral law is going to sit by and say work it out for yourselves what really may be done in this context.

The concept of *partiality* raises similar issues. On that view, morality permits, or even requires, us to prefer our own children over those of others. I don't think so. At least, morality needs to be neutral between agents if it is going to serve as a practical guide for what we as a collection of agents ought to do. Partiality might instruct that I ought to spend $100,000 on piano lessons for my child. Partiality might instruct as well that you ought spend that same $100,000 on a cancer treatment for your child. Partiality might instruct that a town spend that same $100,000 digging new wells in view of the fact that its old ones are a source of the very contaminants that are responsible for that one cancer and many others, existing and future. Partiality, plausibly, might say all these things. Morality, however, is going to tell us which assertion of interest we as a group and we as individuals ought to recognize and respect. To come up with one consistent answer to that question, we cannot assign different weights to the interests that are at stake under the different choices. My child's interest in piano lessons must be weighted the same, when we assess my choice to dedicate the $100,000 to my child's piano lessons, as when we assess your choice to dedicate the $100,000 to your child's cancer treatment. We can't say both that it's obligatory for me to use the money the way I want to use the money but obligatory for you to use the money the way you want to use the money. Again, we face no inconsistency. But if we do not understand morality to be structured in a way that avoids such results, we defeat the purpose of morality if we think that part of the purpose of morality is to identify what we are to do in cases where conflicts arise.

tells us how the tradeoff in wellbeing is to be made in a case in which it is inevitable that at least one person will incur a morally significant loss.

OPPP3 will be useful in many such contexts.

Otherwise Plausible Permissibility Principle 3 (*OPPP3*):
An act *a* performed at a world *w* is permissible *if*

(i) Act *a* imposes a morally significant loss on a person *p* at *w*;
(ii) Each alternative to *a* – that is, each *a′* performed at an accessible world *w′* in lieu of *a* performed at *w* – imposes a morally significant loss on a person *q* (who may but need not be identical to *p*);
(iii) *a* performed at *w* is at least as good for *p* as *a′* is for *q* at *w′*; and
(iv) Either (a) *a* imposes no morally significant loss on any person *r* at *w* other than *p*, or (b) if *a* imposes a morally significant loss on any person *r* at *w* other than *p*, then *a′* imposes a morally significant loss on *r* at *w′* and *a* performed at *w* is at least as good for *r* as *a′* is for *r* at *w′*.

According to OPPP3, the important thing to ask is this. How do the two losses compare? If the loss incurred by the person other than the pregnant woman if the early abortion *is* performed is at least as great and at least as deep as the loss the pregnant woman will incur if the early abortion is *not* performed, then the early abortion is permissible.

OPPP4 also applies to some relatively simple tradeoff scenarios.

Otherwise Plausible Permissibility Principle 4 (*OPPP4*):
An act *a* performed at a world *w* is wrong *if*:

(i) Act *a* imposes a morally significant loss on a person *p* at *w*;
(ii) An act *a′* performed at an accessible world *w′* in lieu of *a* performed at *w* is better for *p* at *w′* than *a* is for *p* at w;
(iii) If a′ performed at *w′* imposes a morally significant loss on any person *q* at *w′*, then that morally significant loss is not as great and not as deep as the morally significant loss that *a* performed at *w* imposes on *p*; and
(iv) Either (a) at most one person – *q* – incurs a morally significant loss under *a′* at *w′* or (b) if *a′* imposes a morally significant loss on any other person *r* at *w′*, then *a′* is at least as good for *r* at *w′* as *a* is for *r* at *w*.

According to OPPP4, the important thing to ask is this. Between the loss the pregnant woman is at risk of incurring and the loss the other person is at risk of incurring, which loss is greater and which cuts deeper? Other things being equal, if one of the two losses is both greater and cuts deeper than the other, then OPPP4 implies that it would be wrong to impose that loss. (OPPP4 leaves open the issue in the case where one loss is much greater but does not cut nearly so deep, such that the person who incurs that greater loss would still be left far better off than the person who incurs the smaller loss. Such is the price we pay to insure that OPPP4 can be agreed upon for inclusion in the otherwise plausible permissibility theory.)

In cases that involve a genuine conflict between the pregnant woman and another person, the facts will often favor the pregnant woman. The loss she will incur if the early abortion is not performed will typically be greater and deeper that any loss that her partner or her grandmother or anyone else can be expected to face. OPPP4 in combination with Variabilism then generates the result that continuing the pregnancy would be wrong.[205]

Nonetheless, we should recognize that in some cases plausible tradeoff principles will imply that the woman is morally obligated to have the baby rather than the early abortion. Such cases, however, may well be so rare that constitutional law in particular should itself unambiguously protect the woman's right to choose.

5.5 Late Abortion

5.4.1 *Late abortion* is the abortion of a fetus that is a *person* and has acquired the capacity to survive as the same person going forward. Allowing the pregnancy to continue until the point at which a first thought signals the coming into existence of a new person is, in effect, *to bring an additional person into existence*. That is so, even though we have let stand the number of *fetal bodies in existence*. Thus, the *fetus* begins to exist when we have a fetus where we once had an embryo. The *person* begins to exist when we have a thinking thing where we once just had a non-thinking fetus.

According to Variabilism, the bare fact that a person has made his or her way into existence at a world means that any loss that that person incurs at that world has full moral significance. That means that any such loss – under any otherwise plausible permissibility theory – *counts against* the choice to create less wellbeing at that world for that person when we could have created more. And one way of creating less wellbeing for that person rather than more would be to truncate that person's existence – to have the late abortion.

Moreover, that person will have had very little chance to accrue much of any wellbeing at all. That person has just begun the process of accruing wellbeing; pleasure, happiness, cognitive activity and all the other things we associate with wellbeing are new to it. If its life is ended *in utero*, the process of accrual will end for it then and there. The loss that we are talking about is, accordingly, very great and would cut very deep. In most cases it will dwarf any loss that the pregnant woman will herself face if the late abortion is not performed.

According to Variabilism, both the loss the pregnant woman will incur if she does not have the late abortion and the loss the developing (in utero) person will incur if the woman does have the late abortion have full moral significance. Both losses *bear on* the permissibility of the late abortion. Let's then suppose that the loss the child will incur if

[205] Again, the objection could be raised that the woman is surely permitted to choose to incur a loss. But see note 204 above.

the late abortion is performed is much greater and cuts much deeper than the loss the pregnant woman will incur if the late abortion is not performed. Then, OPPP3 will itself be blocked: OPPP3 does not generate the result that the late abortion is permissible. OPPP4 is potentially more useful in this context. It is tailored to very simple tradeoff scenarios – cases in which the agent can avoid a very great and very deep loss on behalf of one person by imposing a smaller and shallower morally significant loss on someone else. Thus, when the issue is limited to how the pregnant woman's loss is to be balanced against the developing child's, the facts will ordinarily favor the child. And we will obtain the result that the late abortion is wrong.

5.4.2 There will always be the hypothetical case in which the woman *must* have the late abortion to save the world – or because her own life will otherwise border on the life less than worth having or worse. We can imagine, in other words, cases in which the facts will favor the pregnant woman, and OPPP3 will deem the late abortion permissible or even, in combination with OPPP4, obligatory.

For purposes here, however, the more interesting case is the more routine case – the case in which the gains secured by the late abortion on behalf of the woman or others are less impressive.

Suppose, for example, it is determined, late in pregnancy, that the fetus has a serious genetic or chromosomal abnormality. Suppose that a sonogram performed in the middle of the third trimester reveals a heart defect, and that further testing then reveals Down syndrome.

The anguished pregnant woman wants to have a late abortion. And her reasons for wanting the late abortion may be very good ones. Perhaps she is older and finds the prospect of caring for a seriously impaired child daunting. Perhaps she already has older children with pressing needs of their own that must be met. Perhaps she imagines that, from the Down syndrome child's own perspective, his or her life will count for little – that, while the child's pain will not be excruciating, his or her pleasures will be few.

Let's suppose, however, that in fact it is not reasonable to expect that the life will be anything but well worth living. After all, the child we are imagining has Down syndrome, not the sort of genetic or chromosomal disorder that can reasonably be associated with the genuinely wrongful life, the life less than worth living. Perhaps the child will not have what *the pregnant woman* considers an *ecstatic* life. But still it will be a life well worth living – provided, of course, that the pregnant woman herself *and others* do as they ought for the child going forward. (*Any* life can be made a wrongful life, either through neglect or by intention.)

Is the late abortion permissible in such a case? We suppose again for simplicity that we are dealing with the simplest of tradeoff scenarios, and that no one other than the pregnant woman and the developing child has any interest in whether the late abortion is performed or not. The analysis will then go through as described above – and the late abortion deemed out of bounds by OPPP4 in combination with Variabilism.

5.4.3 The factual element that drives the analysis we have just given is that the loss the developing child will incur at any world where the late abortion is performed will be very great and very deep. It is a loss that, according to Variabilism, has full moral significance. And it is a loss that ordinarily dwarfs any loss that the pregnant woman faces at a world where the late abortion is not performed.

That is so, when the loss the pregnant woman faces derives from the burdens she (or others) will need to take on when it happens that the child she thought would be healthy in fact has Down syndrome or another serious genetic or chromosomal disease or disorder. But it is also so when the loss the pregnant woman faces derives from the fact that her own health has failed in some way. Whether it is her health or even her life that is placed at risk late in pregnancy, in any ordinary case the facts will still favor the developing child.

This last result, however, holds only in the case where the pregnant woman must either endure the full-term pregnancy or have the late abortion. It holds only in the case where there is a genuine conflict. But in many cases there will be no genuine conflict. That is so, since late in the pregnancy the options for the woman are not ordinarily limited to a full-term pregnancy and a late abortion. Instead, an early delivery may well protect the woman's own interests in life and health *without* imposing any very great risk on the developing child.

5.6 Middle Ground on Abortion

I believe that the two main claims just articulated – that early abortion is ordinarily permissible and late abortion ordinarily wrong – represent important middle ground on abortion.

Political conservatives and some religious leaders who disdain abortion at any stage of pregnancy for any reason are surely most concerned to see abortion recognized as impermissible at the later stages. My proposal reflects that concern. It considers late abortions – not just during the ninth month of the pregnancy but any time after the fetus has developed into a thinking thing, a *person*, who has acquired the capacity to survive from one time to another – almost always impermissible. Neither serious fetal abnormality (barring the rare case of the genuinely wrongful life), nor maternal health, nor family circumstances (poverty, or resources stretched thin by the needs of other children), nor the conditions of conception (rape or incest), nor the woman's own ignorance of the fact that she is pregnant or the girl's status as a minor child, would ordinarily be enough to make the late abortion permissible.

By the same token, progressives who think abortion is permissible throughout the pregnancy, including those who consider abortion a matter of bodily autonomy, are surely most concerned to have that fact recognized in connection with early abortion. My proposal reflects their concerns as well. It considers early abortion virtually always perfectly permissible when it is what the woman wants. She need not prove that her fetus is defective, or argue that her health is at stake, or claim or provide evidence that she is the victim of rape or incest. Consistent with the demands of moral law – demands that positive law should be understood to reflect – she may have the abortion for a good reason, a bad reason or no reason at all. She may do just as she pleases, *provided that* she has the abortion within that very generous span of time nature allows between the moment of conception and the moment at which that thing growing within her uterus must be recognized as a *continuing, thinking* thing – the moment, that is, at which the fetus develops from the sort of thing that has no more moral status than a gall bladder into a *person* and the woman's own rough moral equal.

What I am suggesting is that each of the two camps take the ground that is most important to it and concede the ground that is most important to the opposition. It is, I think, no coincidence that that happens to be the recommendation of Variabilism in combination with the otherwise plausible permissibility theory. For Variabilism marks middle ground of a sort, between a view that *radically excludes* merely possible persons from the class of those who matter morally and a view that *radically includes* the merely possible in that same class. Variabilism instead states simply that for each person, merely possible or not, some of that person's losses have moral significance and some do not. From there we then come to the view that some abortions impose morally significant losses and some do not. Where the person already exists, the loss bears on the permissibility of the abortion itself; it *counts against* it. Where the abortion means instead that the person will never exist at all, the loss doesn't bear on the permissibility of abortion; it *cannot* make the otherwise permissible early abortion *wrong*.

5.7 Abortion, Variabilism and the Fourteenth Amendment

5.7.1 There is an argument to be made that a less extreme position on late abortion is in order – that we should just say that late abortion is *highly problematic* but not clearly *wrong*. Of course, we have just said that, according to Variabilism, the loss that the late abortion imposes on the already-existing person has *full moral significance*. But what we have not done is work through the details of the otherwise plausible permissibility theory. It might accordingly seem that we really cannot know what that theory will say about the choice of late abortion in the case where so many other agents are doing things – lawfully, and with what appears to be the full blessing of society – that are *at least* as bad and *getting away with them*.

5.7.2 My own view is that an otherwise plausible permissibility theory will not consider the fact that most other agents are behaving badly a justification for any one agent's behaving badly. That everyone else is doing it may explain from a psychological point of view *why* I've made a choice that is wrong. But it seems that, under any otherwise plausible permissibility theory, it won't make the otherwise wrong act *right*.

5.7.3 Does it, however, make the woman's choice of late abortion *legal*? Liberty is a principle designed to limit the "tyranny of the majority": unless what the individual does harms others, the individual must be left to construct a life that *that individual* takes to have value and impact. But liberty alone is not an adequate constraint on majority tyranny. Equality is important, too.

It could, then, be argued that the bare fact that persons *other than pregnant women* are legally permitted to do very nearly as they please to many human beings and virtually all non-human animals means is that a law requiring *pregnant women* to adhere to a far more stringent standard of conduct violates the equal protection clause of the Fourteenth Amendment of the U.S. Constitution.

Thus, if we can bring veal calves into existence, treat them badly and slaughter them in a way that is frightening and painful to them, if biological fathers can refuse to donate bone marrow to their ailing offspring, if hospitals can deny life-saving medical procedures when their patients don't happen to have the right insurance plan, it may seem unclear that the criminalization of the late-term abortion could itself be constitutional. After all, *not* having the late-term abortion is far more burdensome to the pregnant woman – given the responsibilities that she may well be compelled, morally if not legally, to shoulder once the baby is born – than not eating veal is to me, or donating bone marrow is to the biological father.

Certainly, the question of the morality and the legality of late abortion must be kept strictly apart.[206] Nevertheless, the better view seems to be that the Constitution aligns the cases in just the way that Variabilism suggests. Even if many clear moral transgressions go unpunished by the state, many others are addressed by the state in ways that we think are generally appropriate. The idea that anything goes since not all bad acts have been many illegal or otherwise regulated is not ultimately tenable. If it is correct that the late abortion imposes a loss of the magnitude I have described, and if it is also correct that, as Variabilism suggests, that loss has moral significance, then tight regulation of the late abortion would be constitutionally in order.

At the same time, it seems clear that outright prohibition of the late abortion would take things too far. Laws prohibiting the late abortion, in other words, should include exceptions for the rare case of the genuinely wrongful life and even the perfectly neutral life, as well as for the case where the situation of the pregnant woman is itself so severely limited that the loss she will incur if she does not have the late abortion will be greater and cut deeper than the loss her child will incur if she does.

5.7.4 The constitutional analysis will go very differently for the case of early abortion. There, the moral analysis – at least, the moral analysis that Variability suggests – tracks a very traditional due process analysis, one that protects some of our most important and most personal choices against the vagaries of majority will. The limitation on this due process right of privacy is harm, or loss. The state is understood to have the authority to protect its citizens by regulating a choice that threatens the safety, welfare and education of its people. But it is inconceivable that the Fourteenth Amendment gives states the authority to protect persons *who never exist at all* against the losses they incur at worlds – including *this* world; the *actual* world – where they never exist at all.[207]

It is worth noting that it is constitutionally trivial that the Fourteenth Amendment gives states the authority to protect persons who never exist at all against the losses

[206] See especially Justice Ruth Bader Ginsburg's discussion of abortion in the context of equal protection. Ginsberg (1992), pp. 361–370.

[207] For a contrary view, see Peters (2009).

they would have incurred had we brought them into existences that are avoidably and unjustifiably *bad for them*. That is: a law that had the effect of forcing people to have fewer offspring suffering avoidably flawed or even genuinely wrongful existences would be perfectly constitutional. That is so, even if procreative choice is at stake, and even if the persons who are protected remain, relative to this world, merely possible.[208]

Variabilism's understanding of the *morally significant loss* thus tracks perfectly the concept of *harm* we can expect to see at play within the scope of the due process clause. It's true that the Supreme Court might prefer the term *legally* significant loss rather than *morally* significant loss, but the analysis will be normative all the same.

[208] On this point, Phillip Peters' view is closer to that suggested by Variabilism. See Peters (2009).

Chapter 6
Conclusion

Consider the choice *not* to bring a new person into existence. Suppose that choice creates additional wellbeing for some people but does not destroy wellbeing for any person who does or will exist. A plus for Variabilism is that it enables us to explain why that choice is permissible, or indeed obligatory, *without* washing our hands of that possible person – that *merely* possible person, relative to any world at which the choice is in fact performed – entirely. We can still say that person *matters morally* – that losses that person suffers, at still other worlds where that person does or will exist, bear on the acts that are performed at those other worlds and their alternatives. They *count against* the acts that impose those losses and, in a roundabout way, *in favor of* the acts that avoid them.

In constructing this middle ground, we have plucked one element from a *radically exclusive* approach, according to which none of the losses incurred by any merely possible person have any moral significance at all, and rejected another. And we have plucked one element from a *radically inclusive* approach, according to which all the losses incurred by any merely possible person have full moral significance, and rejected another. And we have been able to construct this middle ground in an organized way, by focusing on *losses* rather than *people*. That is, we have recognized that each person – you, me and the merely possible – incurs some losses that have moral significance and some losses that do not; we each, in other words, *matter morally*, but we matter *variably*. What imbues one loss with moral significance and not another is *where* that loss is incurred in relation to the person who incurs that loss: incurred at a world where the person does or will exist, the loss has full moral significance, and otherwise none at all.

A *practical* plus for this way of looking at the moral status of merely possible persons is that it makes it possible for us to think about abortion without having to have at the table *either* the radical exclusionists *or* the radical inclusionists. What is important, according to Variabilism, is when during the process of the pregnancy the new person comes into existence. For purposes here, I have assumed that *thinking* and *existing*, in the case of persons, come together, where the concept we assign to the term *person* is itself rooted in moral law. This assumption implies, among other things, that a person never exists at all at any world at which that person never has that first thought. And I have suggested that it seems plausible that that first

M.A. Roberts, *Abortion and the Moral Significance of Merely Possible Persons: Finding Middle Ground in Hard Cases*, Philosophy and Medicine 107,
DOI 10.1007/978-90-481-3792-3_6,

thought can be expected to take place at some point during the pregnancy. The embryo does not start out thinking, nor does the early fetus. But the full-term fetus is thinking, in a way that it means it matters morally, perhaps only in virtue of the fact that it matters by then to it how we treat it.

The loss imposed on the possible person by the early abortion – the abortion that by definition takes place prior to that first thought – has no moral significance whatsoever. The early abortion thus will ordinarily be perfectly permissible if it is what the woman wants. That is so, whether the woman wants the early abortion for a good reason, a poor reason or no reason at all.

At a certain point, however, if the pregnancy continues, then we begin dealing with not just a living human organism but also a *new person*, a person whose well-being level will be reduced to almost nothing in the event of the late abortion, a person whose losses will have full moral significance at the world where the pregnancy has continued until thinking itself has commenced. In that case, it's the woman's interests that begin to pale. And the late abortion ordinarily will be wrong.

Here, too, we have constructed a middle ground – taking one element from each of the two sides of the continuing debate and leaving the rest behind. My suggestion is that each of the two camps take the ground that is most important to it and concede the ground that is most important to the opposition.

Appendix A
Otherwise Plausible Permissibility Theory + Variabilism

A.1 OPPP1-OPPP4 + Variabilism

A.1.1 The *otherwise plausible permissibility theory* consists of the four principles OPPP1-OPPP4. Many other principles, no doubt, could be added to the otherwise plausible permissibility theory in addition to those four.

Once the otherwise plausible permissibility theory is paired with a rule that specifies whether the loss a particular act imposes on a person at a world is *morally significant*, that theory can then generate results, for an interesting range of cases, regarding the moral permissibility of the performance of that act at that world. Alternative rules on the moral significance of loss include Inclusion, Exclusion Alpha and Beta and Variabilism. The Neutrality Intuition and the Prior Existence View can also be viewed as providing such a rule.

I call it the *otherwise* plausible permissibility theory to underline that that theory combined with any particular rule on moral significance is plausible and uncontroversial *other than perhaps for* the rule itself, whether that rule is one we have paired with the theory for the purpose of testing the rule or, once the testing is done, for the purpose of applying the theory to a specific issue. The plausible permissibility theory, in other words, stripped of the rule is something we can expect to agree on whether we are Totalists, Averagists or Pluralists (i.e., Inclusionists) or Moral Actualists (Exclusionists) or Person-Based Consequentialists (Variabilists).

The rules themselves rely on the concept of *loss*. For purposes here, the term *loss* is simply shorthand for an individual (or person-based) betterness relation. To say a person p incurs a loss at a world w given an act a relative to a world w' given an alternative act a' is just to say that a' at w' is better for p then a at w. In other words: a person p incurs a *loss* at a world w relative to a world w' *if and only if* agents have created less wellbeing for p at w than they have created for p at w', where w' is itself accessible to those same agents. I take it that the failure to bring a person p into existence at a world w in a case where p has an existence *worth having* – an existence, that is, that is neither *wrongful* nor *perfectly neutral* – under an alternative act a' at an accessible world w' implies that p incurs a *loss* at *w* relative to *w′*.[209] By the same token not bringing *p* into existence at *w* in a case where *p* has an existence *less* than worth having – a wrongful life – at *w′* avoids a loss on p's behalf: *not* bringing p into existence in that case is better for *p*.

Often we compare an act *a* performed at a world *w* against an act a′ performed at a world *w′* for purposes of determining the rightness or wrongness of *a*. It is important to keep in mind that a person *p* can incur a loss at *w* in many different ways. Thus, it's *not just* in virtue of how *p* fares at *w* relative to how *p* fares *at w′* that decides whether *p* incurs a loss at *w*. It's *also* in virtue of how *p* fares *at any third world w″* under an alternative act *a″* that decides whether *p* incurs a loss at *w*. This point is key for analyzing Addition Plus (as well as, e.g., the Mere Addition Paradox).

A.1.2 I begin here by stating Variabilism, since my overall argument is that that loss rule has important advantages over its competitors. I then describe the four principles that, on the basis of the data on moral significance that Variabilism generates, evaluate acts for their permissibility in a range of cases. (Of course, those four principles could also be paired with any competitor to Variabilism, but the permissibility results we then obtain are, I have argued, implausible.)

> *Variabilism*:
>
> A loss incurred at any world *w* by any person *p* has moral significance for purposes of determining the permissibility of any act *a* performed at *w* that imposes that loss and any alternative act *a′* performed at any accessible world *w′* that avoids that loss *if and only if p* does or will exist at *w*.
>
> By implication, the loss incurred by *p* at *w* when agents fail to bring *p* into an existence worth having at *w* will have no moral significance for purposes of determining the permissibility of the act *a* performed at *w* that imposes that loss or any alternative act *a′* performed at *w′* that avoids that loss.

A.1.3 We turn now to the otherwise plausible permissibility theory. OPPP1 is a Pareto principle of sorts. It declares one act wrong when an alternative act makes things better for one person without making them worse for anyone else. But OPPP1 in one respect narrows that idea: the alternative act that proves the one act wrong must itself avoid, not just making things worse for a person than things are under the one act, but also imposing a morally significant loss on any person who does not exist at all under the one act. The exception would be where the person incurs a morally significant loss under each of the two acts but is still better off under the alternative. Thus:

> *Otherwise Plausible Permissibility Principle 1 (OPPP1)*:
> An act *a* performed at a world *w* is wrong *if*

[209] Without the assumption that one can incur a loss by virtue of having been left out of an existence worth having – an assumption that is I think is plausible in any case – our concept of loss would be too narrow to state, test or apply the sort of radically inclusive position that we should be able to associate with Totalism, Averagism and Pluralism. In the absence of that assumption, in other words, what I am call *Inclusion* in combination with the otherwise plausible permissibility theory would not be capable of deeming it ever wrong to leave someone out of an existence worth having, however much that person's wellbeing level contributes to total aggregate wellbeing.

(i) Act a imposes a morally significant loss on a person p at w;
(ii) An act a' at an accessible world w' performed in lieu of a at w is better for p at w' than a is for p at w; and
(iii) Either (a) a' imposes no morally significant loss on any person q at w' other than perhaps p, or (b) if a' imposes a morally significant loss on a person q at w', then a imposes a morally significant loss on q at w, and a' is at least as good for q at w' as a is for q at w.

OPPP2 requires, as a necessary condition on wrongdoing, that the act impose a morally significant loss on some person or another. But OPPP2 leaves open whether a person needs to be an existing or future person at a world in order to incur a morally significant loss at that world. If we add Variabilism to OPPP2, then the necessary condition becomes that an existing or future person incurs a loss. If we add Inclusion instead, then the necessary condition can be satisfied even when it is just a merely possible person who incurs the loss.

Otherwise Plausible Permissibility Principle 2 (*OPPP2*):

An act a performed at a world w is wrong *only if* act a imposes a morally significant loss on a person p at w.

OPPP3 declares an act permissible if each alternative act involves a merely reversing change. Thus:

Otherwise Plausible Permissibility Principle 3 (*OPPP3*):
An act a performed at a world w is permissible *if*

(i) Act a imposes a morally significant loss on a person p at w;
(ii) Each alternative to a – that is, each a' performed at an accessible world w' in lieu of a performed at w – imposes a morally significant loss on a person q (who may but need not be identical to p);
(iii) a performed at w is at least as good for p as a' is for q at w'; and
(iv) Either (a) a imposes no morally significant loss on any person r at w other than p, or (b) if a imposes a morally significant loss on any person r at w other than p, then a' imposes a morally significant loss on r at w' and a performed at w is at least as good for r as a' is for r at w'.

OPPP4 declares an act wrong if it imposes a morally significant loss on one person and agents could have avoided that loss on behalf of that person without imposing a still deeper morally significant loss on another person. Thus:

Otherwise Plausible Permissibility Principle 4 (*OPPP4*):
An act a performed at a world w is wrong *if*

(i) Act a imposes a morally significant loss on a person p at w;
(ii) An act a' performed at an accessible world w' in lieu of a performed at w is better for p at w' than a is for p at w;
(iii) If a′ performed at w' imposes a morally significant loss on any person q at w', then that morally significant loss is not as great and not as deep as the morally significant loss that a performed at w imposes on p; and

(iv) Either (a) at most one person – q – incurs a morally significant loss under a' at w' or (b) if a' imposes a morally significant loss on any other person r at w', then a' is at least as good for r at w' as a is for r at w.

A.2 Extension of Otherwise Plausible Permissibility Theory

OPPP1-OPPP4 are limited for the purposes of constructing a theory that can be used for purposes of testing the competing rules on when losses are morally significant. I elsewhere proposed an approach for how we might go about constructing still other such principles.[210] That approach was developed before I fully understood how useful it is to make explicit the distinctions between Moral Actualism (e.g.) and Variabilism. So it does not use the concept of the morally significant loss. Still, the approach itself may be useful since, at least eventually, we shall want to extend the otherwise plausible permissibility theory.

The approach involves understanding the agents' maximizing tasks as being ordered or structured in a certain way. The agents' first priority is the securing of a better position for the least well-off. Once the position of the least well-off has been secured – once agents have narrowed their options to those that have already been determined not to be wrong by virtue of what they have done to the least well-off – agents may then disregard, for purposes of choosing from among their remaining options, any consequences their choices may have on that particular class of persons. Agents, in other words, can then focus exclusively on the securing of a better position for the next least well-off. And so they may proceed up the ladder – until they may even find themselves, in certain circumstances, required to redress the "plights" of those who are themselves close to being among the very most well-off.

Applying this approach, I assume Variabilism for simplicity. We start with an act a performed at a world w and each alternative act a' performed at each accessible alternative world w'. We identify the persons who are *least well-off* at any such world. We say that those persons occupy the lowest rank – R1 – relative to the entire collection consisting of w and each w' accessible to w. Such persons may have lives less than worth living, or, where wrongful life is not at issue, they may have lives at the zero level, or they may have lives that are worth living. In any different-people case, there will be some worlds that contain, among those persons who do or will exist, at those worlds, persons who do not and will not exist at still other worlds. Each person is assigned a zero wellbeing level at any world where he or she does not and will not exist. That means that, in different-people cases that do not involve wrongful life, we can be sure that R1 consists of all and only persons who have zero wellbeing levels. However, since Variabilism is assumed for purposes here, we can also be sure that anyone in R1 whose wellbeing level is zero in virtue

[210] See Roberts (2002), pp. 344–350.

of the fact that that person never exists at a particular world *w* at all, that person does not incur a morally significant loss at *w*.

Then, we ask whether *a* performed at *w* imposes any moral significant loss on any person who belongs to R1. If it does, then we ask whether there is some alternative act *a′* performed at an accessible world *w′* such that at least some of the persons who (i) occupy R1 at *w* and (ii) incur a morally significant loss at *w* are made better off by a′ at *w′*. Do at least some such persons occupy a higher rank at w′? If so, and if it's not the case that an equal or greater number of people who occupy R1 at *w′* incur morally significant losses at *w′* than the number of people who occupy R1 at *w* and incur morally significant losses at *w* – if, for example, the gain for one person is not achieved on the back of one or more others but is instead achieved *without* imposing a morally significant loss on anyone else *within R1* – then we say that *a* at *w* is wrong. (We need not worry here about whether moving people from R1 to R2 will bring it about that still others are made to endure wellbeing levels that are even lower than R1. By definition, R1 consists of the least well-off; R1 is our base case. Moreover, it may well happen that others up the ladder must incur morally significant losses in order to make things better for persons at R1 who incur morally significant losses. Those persons we *do* need to worry about – but them we worry about later.) We repeat for each alternative act performed at each alternative accessible world *w′*.

But let's go back to *a* at *w*. Let's suppose that it's *not* the case that the criterion described just above for the *impermissibility* of *a* at *w* is satisfied. For all we – yet – know, in other words, *a* at *w* is permissible. We know that, if it's wrong, it is *not* wrong by virtue of anything it does to persons who occupy R1 at *w*. But further scrutiny of *a* at *w* is in order. Accordingly, we repeat the test described above, this time focusing our attention on the effect that *a* has at *w* on persons who occupy R2 at *w*. And, for purposes of applying the test, we now leave out all those alternatives performed at all those accessible worlds *that have already been deemed wrong*.

The test can then be recursively applied all the way through Rn-1, where the most well-off persons themselves occupy Rn. If, for each Rm that is subject to the test, the act *a* at *w* does *not* satisfy the criterion for *impermissibility*, then *a* at *w* should be considered permissible.

Appendix B
The Nonidentity Problem

B.1 Variabilism and the Nonidentity Problem

The nonidentity problem argues that (1) future persons, in many cases, *owe* their very existence to the choices that we make today. Therefore (2) *whatever* choice we make today – however selfish or careless or vile on its face – that choice is not bad for and cannot harm, or impose a loss on, or make things worse for, those future persons (except perhaps in the rare case in which the existence that it accords to them is less than worth having). The nonidentity problem argues as well that (3) many such future-directed choices are themselves clearly wrong. If all of this is correct, then such acts would have to be wrong *notwithstanding* the fact that they are not bad for, and do not harm, or impose a loss on, or make things worse for, any existing or any future person. That would in turn mean that what is called the *person-affecting*, or *person-based, intuition*, according to which any *bad* act must be *bad for* someone, is false.

That result itself seems to compel us in the direction of one of two problematic conclusions. One would be that moral law is, at least in part, *impersonal* in nature. That is, the *bad* act may be *bad for* no one at all but rather bad from the "point of view of the universe." The second is that moral law is not itself fully cogent. The former is the view that *Inclusion* suggests, when it endorses the idea that the losses incurred at worlds by persons who never exist at all at those worlds have moral significance for purposes of evaluating the acts that impose those losses and any alternatives to that act.[211]

Variabilism, put together with the otherwise plausible permissibility theory, offers one way of constructing the person-based intuition. Exclusion Alpha or Beta, or the Neutrality Intuition or the Prior Existence View, have, on occasion, been identified with the person-based approach.[212] In my view, Variabilism offers a far

[211] We can, for example, construe Caspar Hare's "de dicto" response to the nonidentity problem as itself resting on a view that is functionally equivalent to Inclusion as well as R.M. Hare's approach to the problem of abortion. See Caspar Hare (2007), R.M. Hare (1973) and note 36 above.

[212] See Hare (2007).

more attractive interpretation. Isolated inferences we may draw from Exclusion Alpha and Beta, and from the Neutrality Intuition and the Prior Existence View, often seem fine. But these results hardly undo the conceptual and consistency issues that such principles give rise to and that we explored in Chapter 2 above.

At the same time, the *nonidentity problem* is just as much an issue for the Variabilist construction of the person-based intuition as it is for – e.g. – Exclusion Alpha. It is *not* a problem that we can set aside. According to Variabilism, losses incurred at worlds where the subject who incurs those losses never exists at all have *no moral significance*: they cannot make an otherwise permissible act wrong. To judge an act performed at a world to be wrong, we must, according to Variabilism in combination with the otherwise plausible permissibility theory – specifically, OPPP2 – know that that act has imposed a morally significant loss. But that means that act must impose a loss on a person at a world where that person does or will exist. Given that the otherwise plausible permissibility theory is true, the upshot is that, if the person-based intuition is false, so is Variabilism.

B.2 Types of Nonidentity Problems

I believe, however, that the arguments that give rise to the most disturbing types of the nonidentity problem are rooted in fallacy, and that the arguments that give rise to still other types of nonidentity problems do not effectively challenge the person-based intuition. They do not, that is, convince us that the act that maximizes wellbeing for each and every existing or future person at a world is nonetheless wrong at that world.

Thus, I have argued elsewhere that the best first step in resolving the nonidentity problem is to understand that the various cases that make up what is called the *non-identity problem* can themselves be *typed in accordance with their logical features*.

B.3 The Can't-Do-Better Problem

One type of nonidentity problem – we can call it the *can't-do-better* problem – relies on the claim that the act that we take to be wrong and that brings a child into a flawed existence is such that there is nothing agents *could have done* to make things better for that child than they in fact are. The can't-do-better problem relies, that is, on the claim that the flawed existence is in fact *maximizing* for that particular child, and that the only way to undo or mitigate the suffering or burdens of *that* child's existence is simply never to bring *that* child into existence at all.

Examples of the can't-do-better problem include cases in which children are – by way of, perhaps, the treating gynecologist's negligent failure to advise the (soon-to-be) pregnant woman of risk – born with genetic or chromosomal disorders but still have lives that are worth living. Such cases are sometimes discussed under the heading of *wrongful life*, despite the fact that the lives are clearly worth living. *Wrongful disability* is thus perhaps a better term. (When the case involves a *genuinely* wrongful life – a life *less* than worth living – the issue of harm, or loss, is far more straightforward, since agents in that kind of case clearly have a way of making things better for the child. That is, they could have refrained from bringing the child into existence to begin with. Accordingly, the case of the genuinely wrongful life avoids the nonidentity problem from the start. Had the gynecologist exercised due care – had the gynecologist, that is, advised of the value of pre-implantation genetic diagnosis (PGD) or amniocentesis – the woman may well have given birth to a healthier child. But that child would have been *nonidentical* to the child who in fact exists and suffers.

As I have defined to the term *loss* for purposes of this present book – and *harm* elsewhere – it does seem to be the case *the child's coming into existence does not itself constitute a harm or a loss*. But that point will not exhaust the issue of loss. For the gynecologist's negligent act may well mean a scarcity of available family resources. It may mean that the couple has no choice but to distribute scarce resources between the impaired child and any already-existing and future children in a way that *does* impose a loss. It may be impossible, in other words, to protect each existing and future child in the family against the effects of scarcity. If one child is protected against loss, another child may well be forced to incur a loss. All the couple can do, given the gynecologist's negligent act, is choose which child shall incur the loss. The gynecologist's negligent act thus insures that some child or another within the family will incur some loss or another – and a morally significant loss at that.

Actions for wrongful life – or, better, wrongful disability – are thus more accurately described as actions for the compensation, not of the *procreative* effect of the gynecologist's negligence, but rather of the *distributive* effects (which may be incurred by the impaired child *or any of that child's siblings*) of that negligence.[213]

In theory, of course, there will be cases where resources are ample and the disabled child's life is not less than one worth living. There will be cases, in other words, where the gynecologist's negligent act imposes *no* distributive loss and *no* procreative loss and – just to insure we are not missing something – no loss that Variabilism would deem morally significant at all. In other words, the gynecologist's negligence would be maximizing for each person who does or will exist at the world where that negligence takes place. On those facts, OPPP2 instructs that the gynecologist's negligent act is permissible.

In such rarified cases, however, is it clear that that act is *wrong*? An act performed at a world imposes *no morally significant loss whatsoever* – according to Variabilism, and given the account of *loss* we have adopted here – *only if* that act is *maximizing* for each person who does or ever will exist at that world. For my own part, in the unusual case in which that very stringent condition on permissibility happens to be satisfied, it is at least unclear that the choice under scrutiny is itself wrong.[214]

---

[213] I make this argument in Roberts (2009a).

[214] I discuss the issue of wrongful disability and, in general, the can't-do-better problem in more detail in Roberts (2009a), pp. 1–57.

B.4 The Can't-Expect-Better Problem

But the can't-do-better problem is not the most potent type of nonidentity problem. In still other cases, it is incontrovertible that a wrong has been done. At the same time, loss seems elusive.

The argument that global warming cannot harm future generations is a good example of what we can call the *can't-expect-better* problem. So are the arguments against reparations for historical injustices, including slavery and the Holocaust, that rely on the idea that the losses imposed by those acts are limited to their contemporaneous victims, many of whom are now dead, and cannot cogently be understood to be incurred by future generations – the children, grandchildren, and great grandchildren of those contemporaneous victims. So are arguments against harm in the context of Parfit's risky policy and depletion cases, and Kavka's slave child and pleasure pill cases.

In the can't-do-better problem, the argument is that the burden endured by the disabled child is a *physically necessary condition* for that child's ever existing at all. Any alternative act that would have reduced the risk of that child's enduring that condition would have *certainly precluded* that child's coming into existence at all.

In contrast, the can't-expect-better problem relies on the claim that the burdens endured by a particular future person *very probably* will not be lifted without its also obtaining that that person never comes into existence at all. Any alternative act – say, conservation in place of depletion, or vitamin in place of pleasure pill, or feigning to sign the slave child contract in place of validly signing the slave child contract – that would have reduced the risk of that person's enduring that condition would not have *certainly precluded* but would have *vastly reduced the chances* of that person's coming into existence.

From these points we are supposed to infer no harm done and no loss imposed – or at least (and just as good, perhaps, for practical and legal purposes) *very probably* no harm done and no loss imposed.

In fact, however, there is an obstacle to establishing that claim. It is true that it is *highly improbable* that the future person who eventually exists and suffers would have existed, had an alternate and seemingly permissible act – conservation, or vitamin, or feigning to sign – been performed in place of the wrong act in fact performed. But it is a fallacy to think that that point does not hold as well for the wrong act in fact performed. It is a fallacy to think that the wrong act, in some singular way, makes it any *less* improbable that that person will come into existence. The truth is that *that* person's coming into existence is highly improbable whether we perform the wrong act or the seemingly permissible act in its place. Either way, the chances of that particular person's ever coming into existence at all is virtually nil.

This point has its most obvious ramifications in the context of our calculations of *expected value*. I concede that, for any possible future person, the seemingly permissible act will generate very little *expected* value – and that the *actual* value of the wrong act is much greater than that *expected* value for any person who eventually does exist. But a comparison between *actual* against *expected* values cannot establish that one act makes things no worse for a person than another. Any measure of betterness that relies on a combination of *expected* and *actual* values, as we please, is, given the other logical features we want betterness to have, inconsistent.

Nor can a comparison between expected values establish that one act is no worse for a person than another, where the calculation of one of those two expected values *but not the other* is infected by information that we are in a position to acquire only *after* performance. Suppose that act a1 is choosing depletion and then conceiving a child. Suppose that act a2 is choosing conservation and then conceiving a child. If we compare the probability of any particular possible future person *p*'s coming into existence, given a1, against that same probability for that same identical person, given a2, and if, in calculating those probabilities, we limit ourselves just to the information within the grasp of the agent prior to choice, I submit that the two probabilities will be the same, and both will be very close to zero.

The difficulty is that, just as there are too many ways of performing a2 that will take *p* off track for existence altogether, so are there too many ways of performing a1 that will take *p* off track for existence altogether. Prior to choice, we can't know which a2 to choose to bring *p* into existence – but nor can we know which a1 to choose to bring *p* into existence.

I am happy to say that loss, or harm, is a matter of comparing actual value against actual value, or that it is a matter of comparing expected value against expected value. I just think mixing the two is a variation on the principle of *post hoc ergo propter hoc*. And once we recognize that, we can easily avoid the result that a1 is at least as good for *p* as a2 is and obtain that result that a2 is better for *p* than a1 is, and that a1 harms, and imposes a loss, on *p*. This is not to suggest that imposing a *risk* on a person *p* is on its own to impose a loss. After all, in the cases of interest here, we cannot say that, despite the risk, things turned out as well as they possibly could have for the slave child or the victim of depletion. We have – once we set the fallacy aside – *both* risk *and* the fact that that risk has eventuated.

B.6 Conclusion

The upshot, then, is this. The nonidentity problem is really just a collection of different problems displaying distinct logical features. When we fail to distinguish the various nonidentity problems according to their types, we may conclude that the nonidentity problem shows that some "bad" acts are in fact "bad for" no one at all. In contrast, when we examine the problems separately according to their type, the inferences we will draw will be quite different. We can then see that (1) the problems that really do demonstrate that no harm, and no loss, is or will be imposed on any existing or future person – the can't-*do*-better problem – are exactly those in respect of which it never becomes quite clear that a wrong has been done, and (2) the problems that involve acts that are clearly wrong – the can't-*expect*-better problems – are exactly those in respect of which it never becomes even remotely clear that that same person has not been harmed and has not incurred a loss.[215]

---

[215] The final paragraph of Appendix B is based closely on Roberts (2009c), p. 209. I discuss the can't-expect-better problem in more detail in Roberts (2009c), pp. 201–228, and in Roberts (2007), pp. 775–792.

Appendix C
Broome's Inconsistency Argument

John Broome argues that a cogent *theory of goodness* cannot incorporate even the most basic among our so-called *person-affecting*, or *person-based*, intuitions. It's true that he carefully distinguishes *theories of goodness* from *theories of permissibility*.[217] Theories of goodness tell us when one outcome, or distribution, or world, is *morally better* than another. Theories of permissibility tell us which acts are *permissible* and which are not for some range of cases that is of interest to us. He then attests that his objection against person-based theories of goodness does not extend against person-based theories of permissibility. According to Broome, it is "comparatively easy for a [theory of permissibility] to be coherent. It is not constrained by the formal structure of betterness."[218]

Perhaps for Broome it *would* be easy to construct a theory of permissibility that incorporates a person-based approach, addresses an interesting range of cases and is both plausible and cogent. I am not confident that that task is so trivial. But I have a much deeper concern. It is not at all clear to me that theories of permissibility and theories of goodness can be dealt with as entirely separate matters within the scope of moral philosophy. For one thing, we may be unclear what problem a theory of goodness is meant to address if not ultimately the problem of what makes one act permissible and another act wrong. For another, we may object to the idea that – *in virtue of their consequences* – an act *a* can be morally permissible and an act *a′* morally wrong in a case where the outcome that *a′* generates at a given world where *a′* is performed is *morally better* than the outcome that *a* generates at a given world where *a* is performed.

For these reasons, I believe that Broome's objection is potentially a matter of concern whether our interest is in developing a theory of goodness or a theory of permissibility. The purpose of this Appendix C is, accordingly, to take a close look at that objection.

Broome describes four alternative distributions:

A = <3, 1>
B = <2, *>

[217] Broome calls the latter sorts of theories *deontic theories* or, elsewhere, *normative theories*. However, to avoid any hint that what I am proposing here is not fully within the consequentialist tradition, I will call them *theories of permissibility*.

[218] Broome (2009), p. 748.

C = <1, 3>
D = <*, 2>

where the position in the ordered pair indicates the identity of the person affected, and "*" indicates that that person does not exist at all in that distribution.[219]

Using these distributions, we can construct any number of distinct cases. For purposes here, four are important:

Case 1: the case where the alternatives available to the agent(s) are limited to just A and B
Case 2: the case where the alternatives available to the agent(s) are limited to just B and C
Case 3: the case where the alternatives available to the agent(s) are limited to just C and D
Case 4: the case where the alternatives available to the agent(s) are limited to just D and A

I agree with Broome that a *plausible* person-based approach – whether in the form of the four principles Broome attributes to Peter Vallentyne and indirectly to me or in the form of Variabilism in combination with the otherwise plausible permissibility theory – will generate the following results about these four cases:

1. If we are in Case 1, the choice of A is permissible and the choice of B is not.
2. If we are in Case 2, the choice of B is permissible and the choice of C is not.
3. If we are in Case 3, the choice of C is permissible and the choice of D is not.
4. If we are in Case 4, the choice of D is permissible and the choice of A is not.

Moreover, it seems not immediately objectionable to say, on the basis of (1)–(4), that (1′)–(4′) hold as well:

1′. If we are in Case 1, A is morally better than B is.
2′. If we are in Case 2, B is morally better than C is.
3′. If we are in Case 3, C is morally better than D is.
4′. If we are in Case 4, D is morally better than A is.

After all, it is hard to see how (I) one choice could be morally permissible and the other choice morally wrong yet at the same time have it be the case that (II) what is chosen under one choice is exactly as good as, or worse than, what is chosen under the other. At least, it is hard to see how (I) and (II) could both be true given that we are working within a consequentialist framework and – just to make things very easy – given the supposition that the agent happens to know prior to choice what the outcome of that choice will be, so that the choice of A or B or C or D very clearly *is* the choice to bring about (respectively) the distribution A or B or C or D.

I agree, moreover, that if we ignore the fact that our assessments of permissibility and then betterness have taken place within the context of four distinct cases – if, that is, we *detach the antecedents* of (1′)–(4′) – we run into trouble:

1″. A is morally better than B.
2″. B is morally better than C.

[219] Broome (2009), p. 748.

3″. C is morally better than D.
4″. D is morally better than A.

These results, given the transitivity and asymmetry of betterness, lead quickly to contradiction.

But what justifies detaching the antecedents? What justifies the inference from (1′)–(4′) to (1″)–(4″)?

We know that logic alone does not. Broome's own justification is as follows:

> [T]he definition of a distribution of wellbeing ensures that the goodness of a distribution is independent of the choice through which it was made.[220]

It is clearly right that, by definition of *distribution*, the *distribution* itself is independent of the feasible set. The distribution identified as "A," for example, isn't going to change in the sense that wellbeing will be distributed in some distinct way across the population if A is achieved by way of the choice made in Case 4 rather than by way of the choice made in Case 1.

But it doesn't follow from that fact alone that the *goodness* of the distribution is independent of whether the distribution we call "A" is achieved in a case where B is an alternative and D is not rather than in a case where D is an alternative and B is not. It doesn't follow from that fact alone that the *goodness* of the distribution is independent of the feasible set.

Moreover, there is some reason to think that the objection itself is launched on exactly the opposite view. We accepted the move from (1)–(4) to (1′)–(4′) as not immediately objectionable only because we do (roughly) think that to say that A is permissible and B is wrong is to say (among other things) that *A is better than B from a moral point of view*. If somehow we become convinced at the end of the day that we cannot make that kind of connection – that permissibility and betterness are two entirely distinct areas within moral philosophy; or that the former is part of moral philosophy and we don't know what the latter is – then the correct response to Broome's objection would be that we need to revisit the leap from (1)–(4) to (1′)–(4′). We could say that (1)–(4) are perfectly true but that there is nothing in those results that take us to (1′)–(4′).

Broome himself proposes that the advocate of the person-based theory of permissibility take that way out. We can, he suggests, "abandon wellbeing consequentialism" – the idea that "one action is better than another if and only if it leads to a better distribution of wellbeing."[221]

But why should we go down that unpalatable route – that disconnect between permissibility and betterness – when it seems there is a much clearer basis for avoiding the contradiction? Why should we go down that route when we can instead just reject the inference from (1′)–(4′) to (1″)–(4″)? To do that is to insist that the betterness assessments themselves depend on the alternatives available to the agent just prior to choice. A distribution X may be identical to a distribution Y, but X may be exactly as good as Y from the moral point of view in some cases but not in others.[222]

[2220] Broome (2009), p. 751.

[221] Broome (2009), pp. 751–752.

[222] For discussion of whether this approach violates the Axiom of the Independence of Irrelevant alternatives, see Roberts (1998) and Roberts (2003b). See also note 86 above.

Of course, this approach would be opposed by any very classic utilitarian principle, for example, Totalism, a view that, as I construe it, enthusiastically endorses the connection between permissibility and betterness. Totalism ranks all the alternatives in terms of their respective betterness, where one alternative is better than another just in case the summation of individual wellbeing levels under one alternative is greater than that same summation under the other. Then, on the basis of an independent assessment of what the feasible alternatives, relative to a particular agent, world and time, happen to be, Totalism obligates the agent to choose the top-ranked alternative (or from among the top-ranked alternatives in the case of ties). When we do things that way, the immediate detachment of the antecedents is perfectly valid. For it is part of the theory that the betterness ranking is itself independent of what the agent's alternatives in fact happen to be.

But it would beg the question to use this Totalist logic for purposes of justifying the inference from (1′)–(4′) to (1″)–(4″). A person-based approach is going to do things quite differently. Instead of commencing the analysis by ranking the distributions themselves by reference to a *two-place* betterness relation, a person-based approach is going to rank them by reference to a *four-place* betterness relation: X is better for p than Y is for q (where p and q may but need not be identical). Because tradeoffs between distinct persons p and q will sometimes need to be made, and permissibility will depend on whether those tradeoffs are made in a correct way, there will be no single four-place ranking of alternatives on the basis of which any plausible theory of permissibility will instruct the agent to choose the one that is (or among the ones that are) top-ranked. No one is, in other words, to be placed on a moral pedestal, such that the rest of us are to do the best we can *for that one person and that one person alone*.

The four-place relation is itself quite well-behaved. It has all the requisite logical features – transitivity, asymmetry, etc. – and it nicely complies with the Axiom of the Independence of Irrelevant Alternatives. But – if the theory of permissibility itself is person-based in nature – there will be no immediate or direct inference from the top-ranked alternative (or alternatives) to what is permissible, since there will be no guaranty that just because an alternative is highest ranked *for any one particular person* that that alternative is *permissible*.

A more neutral approach will avoid, assuming that we can detach the antecedents and make the inference from (1′)–(4′) to (1″)–(4″). We then avoid Broome's contradiction. Of course, we can easily save the inference if we want to start the discussion by embracing Totalism – or some similarly impersonal approach.[223] But who wants to be a Totalist when it comes to figuring out what we ought to do – or even when it comes to moral betterness? Not Broome, according to Broome.[224]

---

[223] Approaches other than Totalism will also save the inference, including any approach that links betterness strictly to the presence or absence of certain impersonal values that we find displayed in any particular distribution, values that may go well beyond aggregate wellbeing per se, and renounces from the start the connection between betterness and permissibility

[224] Broome (2009), p. 747.

References

Arrhenius, Gustaf. 2003. "The Person-Affecting Restriction, Comparativism, and the Moral Status of Potential People." *Ethics Perspectives* 10(3/4): 185–195.

Arrhenius, Gustaf. 2009. "Can the Person Affecting Restriction Solve the Problems in Population Ethics?" In *Harming Future Persons: Ethics, Genetics and the Nonidentity Problem*, eds. M. Roberts and D. Wasserman. Dordrecht: Springer, pp. 289–314.

Arrhenius, Gustaf. 2010. *Population Ethics*. Oxford: Oxford University Press.

Baker, Lynn Rudder. 2007. *The Metaphysics of Everyday Life: An Essay on Practical Realism*. Cambridge: Cambridge University Press.

Benetar, David. 2006. *Better Never to Have Been: The Harm of Coming into Existence*. Oxford: Clarendon.

Broome, John. 1991. *Weighing Goods*. Oxford: Basil Blackwell.

Broome, John. 1999. *Ethics Out Of Economics*. Cambridge: Cambridge University Press.

Broome, John. 2004. *Weighing Lives*. Oxford: Oxford University Press.

Broome, John. 2009. "Reply to Vallentyne." *Philosophy and Phenomenological Research* 78(3): 747–752.

Bykvist, Krister. 2007. "The Benefits of Coming into Existence." *Philosophical Studies* 135: 335–362.

Carey, Benedict. 2005. "Inside the Injured Brain, Many Kinds of Awareness." *New York Times* (Apr. 5).

Carlson, Eric. 1995. *Consequentialism Reconsidered*. Dordrecht: Kluwer.

Cohen, Cynthia. 1996. " 'Give Me Children or I Shall Die!' New Reproductive Technologies and Harm to Children." *Hastings Center Report* 26(2): 19–27.

Cohen, Cynthia. 2007. *Renewing the Stuff of Life: Stem Cells, Ethics and Public Policy*. New York: Oxford University Press.

Conee, Earl. 1999. "Metaphysics and the Morality of Abortion." *Mind* 108(432): 619–646.

Dasgupta, Partha. 1995. *An Inquiry Into Well-Being and Destitution*. Oxford: Clarendon.

Feinberg, Joel. 1988. "Wrongful Life and the Counterfactual Element in Harming." *Social Philosophy and Policy* 4(1): 145–178.

Feldman, Fred. 1986. *Doing the Best We Can: An Essay in Informal Deontic Logic*. Dordrecht: Reidel.

Feldman, Fred. 1990. "A Simpler Solution to the Paradoxes of Deontic Logic." *Philosophical Perspectives* 4: 309–341.

Feldman, Fred. 1991. "Some Puzzles About the Evil of Death." *The Philosophical Review* 50 (2): 205–227.

Feldman, Fred. 1994. *Confrontations with the Reaper*. Oxford: Oxford University Press.

Feldman, Fred. 2009. *What Is This Thing Called Happiness*? Oxford: Oxford University Press.

Field, Eugene. 1912. "The Duel."

Ginsburg, Ruth Bader. "Sex Equality and the Constitution: The State of the Art." 14 *Women's Rights Law Reporter* 361.

Green, Michael D. 2006. "The Intersection of Factual Causation and Damages." 55 *DePaul Law Review* 671.
Guenin, Louis M. 2008. *The Morality of Embryo Use*. Cambridge: Cambridge University Press.
Hanser, Matthew. 2009. "Harming and Procreating." In *Harming Future Persons*, eds. M.A. Roberts and David Wasserman. Dordrecht: Springer, pp. 179–199.
Hare, Caspar. 2007. "Voices from Another World: Must We Respect the Interests of People Who Do Not, and Will Never, Exist? *Ethics* 117: 498–523.
Hare, R.M. 1973. "Rawls' Theory of Justice." *Philosophical Quarterly* 23(92): 241–252.
Hare, R.M. 1975. "Abortion and the Golden Rule." *Philosophy & Public Affairs* 4(3): 201–222.
Harman, Elizabeth. 2000. "Creation Ethics: The Moral Status of Early Fetuses and the Ethics of Abortion." *Philosophy & Public Affairs* 28 (4): 310–324.
Harman, Elizabeth. 2004. "Can We Harm and Benefit in Creating?" *Philosophical Perspectives* 18: 89–113.
Harman, Elizabeth. 2009. "Harming as Causing Harm." In *Harming Future Persons*, eds. M.A. Roberts and David Wasserman. Dordrecht: Springer, pp. 137–154.
Heyd, David. 1992. *Genethics: Moral Issues in the Creation of People*. Berkeley, CA: University of California Press.
Heyd, David. 2009. "The Intractability of the Nonidentity Problem." In *Harming Future Persons*, eds. M.A. Roberts and David Wasserman. Dordrecht: Springer, pp. 3–25.
Holtug, Nils. 1999. "Utility, Priority and Possible People." *Utilitas* 11: 16–36.
Holtug, Nils. 2001. "On the Value of Coming into Existence." *Journal of Ethics* 5(4): 361–384.
Howard-Snyder, Frances. 2008. "Damned If You Do; Damned If You Don't!" *Philosophia* 36(1): 1–15.
Huemer, Michael. 2008. "In Defence of Repugnance." *Mind* 117(468): 899–933.
Hume, David. 1975/1739–40. *A Treatise of Human Nature*. Ed. L.A. Selby-Bigge. Oxford: Clarendon.
Lee, Susan J. et al. 2005. "Fetal Pain: A Systematic Multidisciplinary Review of the Evidence." *Journal of the American Medical Association* 294(8): 947–954.
Lehrer, Jonah. 2008. "The Eureka Hunt." In *The New Yorker* (July 28), pp. 40–45.
Lewis, David. 1970. "Survival and Identity." In *The Identities of Persons*, ed. Amelie Rorty. Berkeley and Los Angeles, CA: University of California Press, pp. 17–40.
McMahan, Jeff. 1981. "Problems of Population Choice." *Ethics* 92(1): 96–127.
McMahan, Jeff. 2002. *The Ethics of Killing: Problems at the Margins of Life*. Oxford University Press.
McMahan, Jeff. 2006. "Paradoxes of Abortion and Prenatal Injury." *Ethics* 116: 625–55.
McMahan, Jeff. 2009. "Asymmetries in the Morality of Causing People to Exist." In *Harming Future Persons*, eds. M.A. Roberts and David Wasserman. Dordrecht: Springer, pp. 49–68.
McMichael, Alan. 1983. "A Problem for Actualism About Possible Worlds." *The Philosophical Review* 92(1): 49–66.
McNamara, Paul. 2006. "Deontic Logic." *Stanford Encyclopedia of Philosophy*.
Marquis, Don. 1989. "Why Abortion Is Immoral." *Journal of Philosophy* 86: 183–202.
Marquis, Don. 2008. "Abortion and Human Nature." *Journal of Medical Ethics* 34: 422–426
Mill, John Stuart. 2002/1863. *Utilitarianism*. In *The Basic Writings of John Stuart Mill*. Modern Library.
Narveson, Jan. 1976. "Moral Problems of Population." In *Ethics and Population*, ed. Michael Bayles. Cambridge, MA: Schenkman Publishing Co., pp. 59–80.
Parfit, Derek. 1976. "On Doing the Best for Our Children." *Ethics and Population*, ed. M. Bayles. Cambridge, MA: Schenkman Publishing Co., pp. 100–115.
Parfit, Derek. 1987. *Reasons and Persons*. Oxford: Oxford University Press.
Parsons, Josh. 2002. "Axiological Actualism." *Australasian Journal of Philosophy* 80(2): 135–47.
Persson, Ingmar. 2009. "Rights and the Asymmetry Between Creating Good and Bad Lives." In *Harming Future Persons*, eds. M.A. Roberts and David Wasserman. Dordrecht: Springer, pp. 29–47.

Peters, Phillip G. 2009. "Implications of the Nonidentity Problem for State Regulation of Reproductive Liberty." In *Harming Future Persons*, eds. M.A. Roberts and David Wasserman. Dordrecht: Springer, pp. 317–331.

Rawls, John. 1972. *A Theory of Justice*. Oxford: Oxford University Press.

Roberts, M.A. 1998. *Child Versus Childmaker: Future Persons and Present Duties in Ethics and the Law*. Lanham: Rowman & Littlefield.

Roberts, M.A. 2002. "A New Way of Doing the Best that We Can: Person-Based Consequentialism and the Equality Problem." *Ethics* 112: 315–50.

Roberts, M.A. 2003a. "Can It Ever Have Been Better Never to Have Existed At All? Person-Based Consequentialism and a New Repugnant Conclusion." *Journal of Applied Philosophy* 20(2): 159–85.

Roberts, M.A. 2003b. "Is the Person-Affecting Intuition Paradoxical?" *Theory and Decision* 55(1): 1–44.

Roberts, M.A. 2004. "Person-Based Consequentialism and the Procreation Obligation." In *The Repugnant Conclusion: Essays on Population Ethics*, eds. Jesper Ryberg and Torbjörn Tännsjö. Dordrecht: Kluwer, pp. 99–128.

Roberts, M.A. 2006. "Supernumerary Pregnancy, the Problem of Collective Harm and the Nonidentity Problem." *Journal of Law, Medicine & Ethics* 34(4): 776–792.

Roberts, M.A. 2007. "The Nonidentity Fallacy: Harm, Probability and Another Look at Parfit's Depletion Example." *Utilitas* 9: 267–311.

Roberts, M.A. 2009a. "What Is the Wrong of Wrongful Disability? From Chance to Choice to Harm to Persons." *Law and Philosophy* 28(1): 1–57.

Roberts, M.A. 2009b. *Harming Future Persons: Ethics, Genetics and the Nonidentity Problem*, ed. with David Wasserman. Dordrecht: Springer.

Roberts, M.A. 2009c. "The Nonidentity Problem and the Two Envelope Problem: When Is One Act Better for a Person than Another?" In *Harming Future Persons: Ethics, Genetics and the Nonidentity Problem*, eds. M. A. Roberts and David Wasserman. Dordrecht: Springer, pp.201–228.

Rosen, Gideon. 1990. "Modal Fictionalism." *Mind* 99(395): 327–354.

Russell, Bertrand. 1903. *Principles of Mathematics*. New York: Norton (reprinted under Berne Convention).

Sen, Amartya. 1992. *Inequality Reexamined*. Cambridge, MA: Harvard University Press.

Sher, George. 1977. "Hare, Abortion, and the Golden Rule." *Philosophy & Public Affairs* 6(2): 185–190.

Sider, Ted. 1993. "Asymmetry and Self-Sacrifice." *Philosophical Studies* 70(2): 117–132.

Sider, Ted. 2002. "The Ersatz Pluriverse." *Journal of Philosophy* 99: 279–315.

Sidgwick, Henry. 2006/1884. *The Methods of Ethics*, 7th ed. Elibron/London: Macmillan.

Singer, Peter. 1976. "A Utilitarian Population Principle." In *Ethics and Population*, ed. M. Bayles. Cambridge, MA: Schenkman Publishing Co., pp. 81–99.

Singer, Peter. 1994. *Rethinking Life and Death: The Collapse of Our Traditional Ethics*. St. Martin's Griffin.

Singer, Peter. 1999. *Practical Ethics*, 2nd ed. Cambridge University Press.

Steinbock, Bonnie. 2009. In *Harming Future Persons: Ethics, Genetics and the Nonidentity Problem*, eds. M. Roberts and D. Wasserman. Dordrecht: Springer, pp. 155–178.

Taddio, Anna et al. 1997. "Effect of Neonatal Circumcision on Pain Response During Subsequent Routine Vaccination." *The Lancet* 349(9052): 500–603.

Taleb, N.T. 2007. *The Black Swan: The Impact of the Highly Improbable*. New York: Random House.

Tännsjö, Torbjörn. 2004. "Why We Ought to Accept the Repugnant Conclusion." In *The Repugnant Conclusion: Essays on Population Ethics*, eds. Jesper Ryberg and Torbjörn Tännsjö. Dordrecht: Kluwer, pp. 219–237.

Temkin, Larry. 1993. *Inequality*. New York: Oxford University Press.

Thomson, Judith Jarvis. 1971. "A Defense of Abortion." *Philosophy and Public Affairs* 1(1): 47–66.

Vallentyne, Peter. 2000. "Equality, Efficiency, and the Priority of the Worse-Off." *Economics and Philosophy* 16: 1–19.
Velleman, David. 2008. "Persons in Prospect" *Philosophy & Public Affairs* 36(3): 221–322.
Wasserman, David. 2008. "Comment on Hare." *Ethics* 118(3): 529–535.
Weinberg, Rivka. "Existence: Who Needs It?" Unpublished Manuscript.
Wolf, Clark. 1997. "Person-Affecting Utilitarianism and Population Policy; Or, Sissy Jupe's Theory of Social Choice." In *Contingent Future Persons*, eds. N. Fotion and J.C. Heller. Dordrecht: Kluwer, pp. 99–122.

Index

The manufacturer's authorised representative in the EU is Springer Nature Customer Service Centre GmbH, Europaplatz 3, 69115 Heidelberg, Germany. If you have any concerns regarding our products, please contact ProductSafety@springernature.com

Printed and bound by CPI Group (UK) Ltd, Croydon, CR0 4YY
25/09/2026
02222898-0002